SALMONELLA

Rufus K. Guthrie

CRC Press
Boca Raton Ann Arbor London

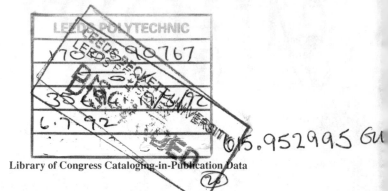

Library of Congress Cataloging-in-Publication Data

Guthrie, Rufus K.
    Salmonella / author, Rufus K. Guthrie
        p.   cm.
    Includes bibliographical references and index.
    ISBN 0-8493-5419-6
    1. Salmonellosis. I. Title
    [DNLM: 1. Salmonella. 2. Salmonella Infections.   WC 269 G984s]
QR201.S25G88 1991
615.9′52995—dc20
DNLM/DLC   91-19363
for Library of Congress   CIP

This book represents information obtained from authentic and highly regarded sources. Reprinted material is quoted with permission, and sources are indicated. A wide variety of references are listed. Every reasonable effort has been made to give reliable data and information, but the author and the publisher cannot assume responsibility for the validity of all materials or for the consequences of their use.

Direct all inquiries to CRC Press, Inc., 2000 Corporate Blvd., N.W., Boca Raton, Florida, 33431.

© 1992 by CRC Press, Inc.

International Standard Book Number 0-8493-5419-6 ✓

Library of Congress Card Number 91-19363
Printed in the United States 0 1 2 3 4 5 6 7 8 9

PREFACE

Salmonella infections of the human historically were largely limited to infections caused by *Salmonella typhi*, or typhoid fever. In much of the world, that infection, typhoid fever or enteric fever, also included concerns about the paratyphoid infections. With treatment of water supplies and improvement in disposal of human waste, concern with all of these infections has decreased somewhat in this country.

Unfortunately, with the decreased incidence of typhoid and other enteric fevers in the U.S. and other developed nations, there has not been a corresponding decrease in either incidence or concern with *Salmonella* gastroenteritis. Rather the reverse has been true, and there has been increased incidence and concern with salmonellosis.

Salmonellosis was hardly considered as a potential waterborne disease before 1940, but in 1976 to 1980, 6 waterborne outbreaks were identified, with a total of over 1000 cases. In 1985, there were over 60,000 diagnosed cases of salmonella infections other than typhoid in the U.S.; an increase from no cases reported to the Center for Communicable Diseases in 1945.

Not surprisingly then, an increased concern on the part of food industries, the public, the media, and regulatory agencies from the U.S. to the World Health Organization has developed in recent years. This book was written to provide some background of understanding of the infection, how the causative organisms change in and are affected by the environment, some effective measures for controlling spread of the pathogens, and how the causative organisms are spread through the environment to reach the human in sufficient numbers to provide an infectious dose. If there is some understanding of these things on the part of the reader, and on the part of persons in the food industries, then the efforts to control spread of *Salmonella* sp. should be better understood and perhaps such control efforts will be made more frequently and more effectively.

THE AUTHOR

Rufus K. Guthrie, Ph.D., is professor of Microbiology and Human Ecology, Department of Disease Control, the University of Texas, School of Public Health, Houston, TX.

Dr. Guthrie graduated in 1948 from the University of Texas at Austin with a B.A. degree in bacteriology and obtained a M.A. degree in 1950 from the University of Texas at Austin. In 1954, he received his Ph.D. degree from Baylor University, College of Medicine, Houston, TX, in microbiology and biochemistry.

Dr. Guthrie is a member of the American Society for Microbiology; Texas Branch, ASM; Sigma Xi; Texas Academy of Sciences, Fellow; American Public Health Association; Texas Public Health Association; International Biodeterioration Society and a Charter Member of the Pan-American Biodeterioration Society; and a consultant for the American Council on Science and Health.

Among other honors, Dr. Guthrie was elected Fellow in the American Academy of Microbiology.

Dr. Guthrie has presented 5 invited lectures at international meetings, 10 invited lectures at national meetings, and approximately 20 guest lectures at universities and institutes. He has published more than 150 research papers. His current major research interests are in infectious disease transmission, including salmonellosis, vibriosis, and campylobacteriosis.

*Dedicated
to
Maxeen*

ACKNOWLEDGMENTS

This book is dedicated to my wife whose support and encouragement through the years has made my work meaningful. I deeply appreciate the love and encouragement she and our children have shown in all my endeavors. I must also express my thanks for the computer drawings, computer assistance, computer teaching, patience, friendship, and help from Garth Morgan. The patience, photocopying, collating, and general help of Peggy Donnellan has made this writing, my teaching, and my work in general much easier. The help of Kendra Freels and Olaf Haerens has made this project and many others go much smoother.

TABLE OF CONTENTS

Chapter 1

THE *SALMONELLA*

"*Salmonella* in eggs! Botulism from garlic-in-oil! *Listeria* in cheese! It seems that every day newspapers are shouting headlines about outbreaks of foodborne illnesses. Just within the past year, for example, eggs contaminated with *Salmonella* bacteria have sickened scores of consumers in the northeastern U.S., airline food tainted with *Shigella* bacteria brought down members of the Minnesota Vikings football team, and three residents of New York state were hospitalized with botulism after they ate an unrefrigerated garlic-in-oil mix. In fact, the Food and Drug Administration scientists estimate that tens of millions of cases of foodborne disease occur every year in this country." So begins an article by Catherine Carey writing for the *FDA Consumer* in June 1989.[1]

Infections other than salmonellosis are involved in the statement by Catherine Carey, but this statement serves to emphasize the importance of foodborne disease. That importance is similar to the importance of salmonellosis. According to numbers of cases reported and in the opinion of many in the public health professions, salmonellosis is one of the most common (if not the most common) infectious diseases transmitted by contaminated foods. The most feared infection caused by bacteria in this genus is typhoid fever, an infection caused by *Salmonella typhi;* most feared because it is by far the most severe of the salmonellosis infections in terms of symptoms and outcomes.[2] Although most feared, the incidence of typhoid fever is not nearly so great as the incidence of other infections caused by other *Salmonella* serovars which are called salmonellosis. It has also become true in more recent years that those cases of typhoid fever which do occur, in this and other developed countries, are most likely to be foodborne rather than waterborne as in the past.[2] Changes in the pattern of incidence of salmonellosis have been noted in both the U.S. and U.K. since the days of World War II in the 1940s. Black et al.[3] suggest that the actual incidence of salmonellosis in this country is not truly measured by the system which has been in place since the 1940s. He gives as an example the fact that in the years 1953 to 1955, the reports of outbreaks of infection in England and Wales were 28 times as high as were reports in the U.S. One of the reasons given for the belief that our reports were too low was the fact that in Great Britain, a cause of their high incidence of salmonellosis was believed to be the importation of powdered eggs from this country. Those authors also express the belief that an increased use of mass food processing and distribution was a factor in the increase in incidence of salmonellosis, as were advances or changes in surgical and medical procedures which made the risk of infection by *Salmonella* higher and with the increased risk. There were also increased economic costs to the public in general.

TABLE 1.1

Ten Most Important Infectious Disease Groups According to the National Foundation for Infectious Diseases — Ranked as Priorities for Research

Top Ten Infectious Diseases, 1988

AIDS
Hospital-acquired infections
Hepatitis
Diarrheal diseases
Meningitis and encephalitis
Respiratory infections
Sexually-transmitted diseases
Infections in cancer and
 transplantation
Urinary tract infections
Tropical and parasitic diseases

From National Foundation for Infectious Diseases, T*he Double Helix,* 3(3), 3–5, 1988. With permission.

The National Foundation for Infectious Diseases[4,5] recently revised the rankings of the top ten infectious diseases in order to set priorities for research funding, training, and the need for preventive education (Table 1.1). This ranking is not in order of numbers of cases, but is in general arranged according to the impact expected on public health in order to set research funding priorities. Fourth among these diseases was the general classification of "diarrheal diseases". This includes many infectious diseases, among them many of which are transmitted by contaminated foods, and certainly includes salmonellosis.

Black et al.[3] from Massachusetts General Hospital reviewed unusual aspects of the cases of salmonellosis over a period of 6 years. They concluded that salmonellosis can be divided into four different clinical syndromes: gastroenteritis, bacteremia with or without extraintestinal localization, enteric fever (typhoid-like syndrome), and the carrier state (either temporary or permanent). Of these, their findings indicated that gastroenteritis was the syndrome appearing in approximately 70% of the cases, and that salmonellosis most often occurred following consumption of contaminated food — regardless of the clinical syndrome which appeared. In these cases, *S. typhi* always caused an enteric fever type of disease, and the unusual infections were always caused by some other serovar or strain of *Salmonella*. On the other hand, the carrier condition rarely occurred other than in those patients infected with *S. typhi*. The temporary carrier condition (convalescent carrier) occurred in other infections, but very rarely became permanent. The investigators made a point of the fact that symptomless carriers are an important source of the organisms when these persons are occupied as handlers of foods. At least a number of such carriers probably acquired the condition while employed as handlers of uncooked meats and carcasses of animals to be used for foods. Their findings were that of 500 outbreaks in the U.S., the Caribbean area, and South America, 56% of the outbreaks were traced to symptomless carriers employed as food handlers.[3]

It is generally believed that only a very small portion of the cases which do occur are ever recognized, and it has been said that as few as 1% of the clinical cases of salmonellosis are reported to the proper authorities.[6] Todd[7] calculated that for Canadian and U.S. data, the numbers of reported cases need to be multiplied by a factor of 350 to approach what is likely to be the actual number of cases occurring. There is general agreement that salmonellosis, or salmonella gastroenteritis occurs frequently in small outbreaks as foodborne infections, and that children and the elderly are the most likely targets for this infection. Larger outbreaks are likely to occur in hospitals, other institutions, restaurants, or nursing homes. Hospital outbreaks are likely to last for some considerable period of time since the organisms tend to survive in the environment, and are cross-transmitted by personnel who handle contaminated food, clothing, or instruments.[6] It has been estimated that there may be 2 to 3 million cases of salmonellosis in the U.S. each year, with only a small proportion of these appearing as a part of the statistics of the disease.

Reporting outbreaks of salmonellosis and other foodborne infections helps keep the public aware of the constant danger of this type of infectious disease. Although the fatality rate from most of these infections is comparatively low, the fact that there are deaths occurring, and that there are consistent increases in numbers of cases reported, continue to remind that these infections are important factors in the total picture of public health and infectious diseases in this country, and indeed worldwide. The incidence and the fatality rates of these infections remain higher in the underdeveloped than in the developed nations of the world. The poor reporting in this country, and the less efficient reporting of such infections in other parts of the world, may lead to a false sense of security in all countries, but we need to be reminded on occasion that these diseases do still exist worldwide, and that the death rates from these infections are higher in other countries than in the U.S. As long as the diseases exist anywhere in the world, they can be transported, and transmitted to persons in any country by populations as mobile as now exist.

Although reporting of infectious disease occurrence in this country is likely to be more accurate than in many others, it is apparent from our history that our reports are not, and never have been accurate. In 1893 an Act of Congress authorized collection of information on the occurrence of infectious diseases from state and municipal offices on a weekly basis. Since that time, a gradual increase has occurred in the number of reports received by the Public Health Service, but not all states began to report regularly until 1925. The Communicable

Disease Center was responsible for the reporting of venereal disease in 1957, for tuberculosis in 1960, and, finally, for nationally reported diseases in 1961. Changes in the characteristics and occurrence of diseases over the years have required that the reporting system be frequently modified, and that new diseases be periodically added.

Examination of the economic effects of outbreaks of salmonellosis in the human reveals that this impact may be considerable. A recent report[8] indicates that in developing countries with economies which are marginal, the impact of foodborne diseases such as salmonellosis may be devastating. Another expert committee[9] analyzed the different factors which go into calculations of the costs of outbreaks of foodborne diseases. That report concluded that patients must include payment for hospitalization, physician services, nurse services, medications, work time lost, pay lost, microbiological testing, and follow-up costs. The outbreak also incurs expenses for workers compensation, food industry costs to locate the problem and correct it (this may involve lay-off time for workers and rearrangement of work schedules), and all aspects of investigation into the source and nature of the infection spread.

Todd[7,10] has written two excellent summaries of the numbers of cases of infections (including resulting deaths) and the costs of foodborne diseases in the U.S. and Canada. In those articles, it is obvious that in both countries, *Salmonella* (other than typhi) infection is the single most common foodborne disease. Figure 1.1 depicts the changes in numbers of all reported cases of salmonellosis from 1941 to 1985. Although accurate numbers of infections are not available, the estimates in those reports were that there were 2,960,000 cases of infection resulting in 31.9 deaths in the U.S. and that there were 625,408 cases of infection resulting in 8.1 deaths in Canada. These numbers were presented as means for the years 1978 to 1982. The numbers for *S. typhi* infections were much smaller; however, the infection did still exist in the U.S. although no outbreaks occurred in Canada during this period of time. In the U.S., the fatality rate for *Salmonella* infections other than *S. typhi* was reported as 0.001, whereas the fatality rate for *S. typhi* infections was reported as 0.05. From these figures, it is obvious that the *S. typhi* infections are much more serious and life threatening than are other salmonelloses.

Todd[7,10] estimated that there are a total of about 12.5 million cases of foodborne infection in the U.S. each year, and that these cost the American public about $8,426,000,000 each year. Since salmonellosis, excluding typhoid fever, accounts for almost half of this figure, the impact of these infections on public health and the infectious disease picture in this country is obvious.

Todd[7,10] also reported that the average cost per case for salmonellosis infection in the human in the U.S. was $1350, as it was in Canada. The total cost of salmonellosis infections in the U.S. was calculated at $3,991,000,000 and in Canada at $846,200,000, excluding the costs of typhoid infections in the U.S. and not including the costs of salmonellosis in poultry in Canada as is shown in Table 1.2. In the U.S., Todd estimated that the costs of those cases of

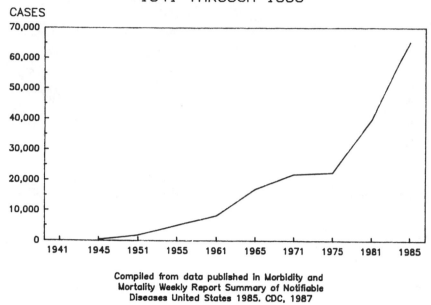

CHANGES IN NUMBER OF REPORTED CASES OF SALMONELLOSIS NOT INCLUDING TYPHOID FROM 1941 THROUGH 1985

Compiled from data published in Morbidity and
Mortality Weekly Report Summary of Notifiable
Diseases United States 1985. CDC, 1987

FIGURE 1.1   Change in number of reported cases of salmonellosis not including typhoid from 1941 through 1985. Graph courtesy of Garth Morgan.

salmonellosis resulting from mishandling at the food service step was $2,494,000,000, and the costs of those resulting from mishandling in food processing was $1,486,000,000. These numbers give a good indication of the relative importance of food contamination during these steps in making food available for consumption; the indication being that contamination during food service handling is much more likely to cause infection in the consumer than in most food processing steps. This is no doubt due to the fact that, in most cases, after food processing in commercial establishments, the food product, either immediately or at least before it is consumed, is again treated to negate contaminants present and, therefore does not as often result in infection of the consumer.

Other cost estimates were reported for specific outbreaks in this country in addition to the illness, suffering, and in some cases deaths, resulting from salmonellosis outbreaks. There are always economic costs to be calculated resulting from these illnesses. These economic costs were calculated and reported for an outbreak involving 125 cases of infection occurring in the early 1970s.[11] This incident occurred as a result of cross-contamination of foods as they were being prepared in the restaurant kitchen. There were no deaths, but

TABLE 1.2
Cost-Benefit Comparison of *Salmonella* Control Measures in the Canadian Poultry Industry (1987)

Control measure	Cost (millions of Can. $)	Benefit (millions of Can. $)[a]	Other benefits
Cleaning of hatching eggs and hatcheries (maintaining hygenic conditions and fumigating eggs)	3.9	1.3	Reduction in other poultry diseases
Production of clean feed-stuffs (fumigation in feed mills and rendering industry)	7.5	9.8	Reduction in other poultry diseases
Cleaning of grower barns	7.6	8.6	Reduction in other poultry diseases
Nurmi culture[b]	25.7	24.0	
Cleaning of poultry crates (cleaning and disinfection, efficient crate washers)	0.8	13.2	Reduction in other bacteria
Clean poultry processing	0.6	8.4	Reduction in other bacteria
Irradiation of processed poultry	18.5	52.7	Reduction in other pathogens, increased shelf-life
Education of the consumer (once per year)	0.4	7.4	Reduction in other bacteria and in salmonellan contamination of other foods
Education of employees in the food service industry (every 2 years)	1.9	10.1	Reduction in other bacteria and in salmonella contamination of other foods

[a] The benefits refer to the reduction in human salmonellosis and associated costs and/or an increase in the productivity in the poultry sector
[b] The Nurmi culture is not yet used on a commercial scale.

From Salmonellosis Control: The Role of Animal and Product Hygiene, Tech. Rep. Ser. No. 774, World Health Organization, Geneva, 1988. With permission.

there were 11 hospitalizations among the 50 who consulted physicians. The costs which were calculated included lost salaries and productivity of those ill, medical and hospital expenses, and costs of investigation of the outbreak as well as the consequences of the outbreak on the owner of the restaurant.

The author estimated that 64% of the cost of the outbreak was represented in the $18,413 in lost salaries and productivity of the persons who became ill. This total loss was derived from 94 persons, since the remainder were not employed outside the home at the time of illness, and therefore, the productivity losses were not included in the study.

Medical and hospital expenses, considerably less in the 1970s than currently, were estimated at $2965 or 10.3% of the total cost of the outbreak. Certainly this category of cost would amount to a much larger portion of the total in today's medical environment. The cost of investigating the outbreak was calculated as $2355 or 8.2% of the total. This category would also be found to comprise a larger portion of the total today. The authors acknowledged that the cost to the restaurant owner could not be accurately determined and arbitrarily assessed an amount of $5000 for this item, a figure amounting to 17.4% of the total. It was pointed out that at the time this article was written, in one state alone (Minnesota), approximately five outbreaks of this size occur each year. If each of those cost the same as this one ($28,733), then the total for the year in this state would be $142,665 at those price levels and would certainly be higher today. At the time of that article, the authors also calculated that the cost of inspecting a food service establishment was $10.70. Assuming that the cost of inspection has increased at the same rate that other costs have gone up, this would allow inspection of 14,277 food service establishments for the cost of this one outbreak.

Salmonellosis is still considered to be the number one cause of gastroenteritis in the world, but some authorities are beginning to question this assertion in view of the number of outbreaks and cases of campylobacteriosis which have been reported in recent years. The above numbers of cases and costs in just two countries give a good indication that this is true in regard to salmonellosis. The number of cases of campylobacteriosis actually culturally diagnosed has not come close to approaching the numbers of cases of salmonellosis actually diagnosed, and many estimates do not equate the two. Much more accurate data is needed to ascertain the true importance of campylobacteriosis as a foodborne infectious disease. Fortunately, the procedures which should help to reduce or control salmonellosis should also help to reduce or control campylobacteriosis.

In recent years, the infection of greatest concern in the U.S. has been infection caused by *Salmonella enteritidis* which has occurred in many outbreaks, particularly frequent in the Northeastern region of the U.S. For example, in 1989, there were more than 49 outbreaks caused by *S. enteritidis* in nine states and Puerto Rico.[12] These outbreaks involved 1628 cases resulting in 13 deaths. Examples of these outbreaks are found in such cases as a baby shower in New York, after which 21 persons became ill after eating a pasta dish made with a raw egg. One of these cases was a female who was 38 weeks pregnant. She delivered while still ill, and the infant developed *S. enteritidis* blood poisoning and required lengthy hospitalization. In another outbreak, 15 patients were hospitalized, and one 49-year-old male died after eating a custard pie made with eggs contaminated with *S. enteritidis* at a company party in Pennsylvania. If one looks at several years, from January 1985 through October 1989, there were 189 outbreaks caused by *S. enteritidis* with 6604 cases and 43 deaths. Dr. Joseph Madden of the FDA was quoted as saying that probably many more illnesses caused by this organism went unreported in this period.[12]

8   *Salmonella*

TABLE 1.3
Carriers vs. Cases of Typhoid Reported in the U.S.

	1975	1980	1984
Typhoid cases	375	510	390
Typhoid carriers	NA	62	54
Cases/100,000 population	0.18	0.23	0.17
Carriers/100,000 population	NA	0.03	0.03

From Annual Summary 1984, Morbidity and Mortality Weekly Report, Centers for Disease Control, Public Health Service Atlanta, GA, 1986.

In the past, outbreaks of salmonellosis which have attracted the most attention and caused the most fear have frequently been those involving *S. typhi,* and typhoid fever. The number of cases of typhoid fever in the U.S. in 1952 had fallen to below 1000 diagnoses per year.[13] One particularly noteworthy series of incidents, all involving typhoid and all involving the same mode of transmission and method of maintenance in the environment is found in the story of "Typhoid Mary", without doubt the most notorious carrier ever identified with the transmission of this infection.

A carrier is a healthy person who harbors and spreads pathogenic organisms without showing any signs or symptoms of an infection. Poor hygienic practices of the individual who is the carrier aids in transmission of the organism to susceptible persons, either through vehicles, such as food, or by direct contact. The carrier condition in the case of typhoid, and of other salmonelloses, is relatively common and may be described as chronic or permanent. It is cured only with much difficulty if at all.[2]

In the 1940s the carrier rate was variously estimated in the U.S. at 1:2500 to 1:3500 and upward. The numbers of typhoid cases as compared to the numbers of carriers diagnosed are shown in Table 1.3. In 1961, the carrier rate in England was estimated to be as low as 1:100,000. The Centers for Disease Control have published the carrier rate in 1984[14] in the U.S. as being 0.03:100,000; which is identical to the rate reported for 1980.[14,15] These rates are probably low because so many carriers are known to be intermittent shedders of the organisms. In intermittent or light excreters, it is most difficult to detect the positive carrier condition. The carrier state also does not always correlate with the disease rate as is observed in diphtheria. Diphtheria infection is relatively rare in tropical or hot climates; however, the carrier rate in those areas is found to be approximately the same as in temperate zones where the disease occurs more frequently.[2]

A report of the "Typhoid Mary" case written by Major George A. Soper, one of the actual investigators of the case, makes fascinating reading for persons interested in infectious disease and the development of the science of epidemiology.[16] This story emphasizes the potential for spread of any salmonellosis infection by a single carrier who has poor hygienic habits, who works in the food industry, and is shedding the pathogenic organisms on a fairly regular basis. Mary was a cook in the New York area who worked for a number of different households over a period spanning 15 years. In many of the homes where Mary worked, one or more members of the household was later diagnosed as having typhoid fever. The infections which occurred were usually not in family members, who rather consistently were wealthy, because the foods were subjected to heat after Mary, the cook, had come in contact with them. The infections usually occurred in the servants of the household who were served directly by the cook, or in family members who were served directly by Mary and were served foods not subjected to heat after being handled by the cook. In each outbreak of the disease, the infection occurred after Mary had moved on to a new location, and except for the intensive and relentless investigative work of Major Soper, the real cause of the outbreaks of infection would have likely not been discovered. Each time Mary moved and left behind the tell-tale cases of typhoid among those who had consumed her cooking, Major Soper was eventually able to track her, and finally was able to force the issue so that Mary was identified as a carrier of the typhoid bacillus. As was possible in those times, Mary was barred from employment in a food handling capacity after she was released from incarceration for the diagnosis of her carrier condition. Although she agreed not to seek employment in a food handling capacity, she broke her agreement after her release, and as stated by Major Soper: "She knew the danger and how to avoid it. She knew that she was violating her agreement with the Department of Health in engaging in the occupation of cook. That she took chances both with the lives of other people and with her own prospect for liberty and that she did this deliberately and in a hospital where the risk of detection and severe punishment were particularly great, argues a mental attitude which is difficult to explain." Unfortunately, even when aware of the carrier condition, as seen in this case, we can not depend upon the voluntary compliance of persons shedding these organisms. All too often, the carrier may not be aware that the condition exists, and therefore, can not take measures to protect contacts. The presence of unknown carriers in the general population provides a potential for contamination of foods at any stage of production, processing, and preparation. If the contamination occurs at a time when the food will not be treated by heat or other measures prior to consumption, the transmission of the organisms, and therefore, of the infection, becomes a possibility. The one-person epidemic in the New York area around 1900 which was caused by "Typhoid Mary" covered a time span of over 15 years and resulted in a least 51 cases of infection (that number was diagnosed). Any one of the persons infected could have become a carrier, thus creating the potential for a much larger scale epidemic of the disease than that which actually occurred.

The importance of the story of "Typhoid Mary" lies not in the number of cases of infection which resulted from her activities, but rather in the demonstration that over a period of 15 years the carrier state served to continue spread of infectious disease from one location, and susceptible population, to another. Although the carrier condition has been demonstrated to be not as important to other salmonellosis infections as it is in typhoid fever,[13] the potential for the establishment of a carrier condition in any intestinal infection is always present, and this situation must be remembered in program plans for controlling these diseases.

There have been outbreaks of salmonellosis of considerable size reported in the time span since the discovery of the cause of this disease; however, the largest outbreak of the *Salmonella* infection from one food source occurred in the Midwest in the U.S. in 1985. Over 16,000 persons were known to have been infected in that outbreak spread by milk processed in a single dairy plant in Illinois.[17] The organism responsible was a strain of *S. typhimurium*. This particular strain is recognized to be resistant to certain antibiotics that might have helped to control infections if they had been caused by other strains of the bacteria. Individuals infected from the milk complained of diarrhea, fever, abdominal pain, and cramps, and no doubt most of the consumers of the milk recovered quickly and never consulted a physician.

In this outbreak, one of the more surprising aspects was that milk was involved. The dairy industry has for many years had one of the more enviable records in the food industry as far as safety of products is concerned. This record is no doubt in part due to the fact that dairy products in this country are usually pasteurized, and the pasteurization process kills most contaminating bacteria which are pathogenic for the consumer. Unpasteurized milk, with some frequency, may contain *Salmonella, Campylobacter, Listeria,* or other pathogens. In the case of this salmonellosis outbreak, the process broke down because of a practice of postpasteurization blending to produce low fat milk for sale. This is another example of the fact that each time a step is added to a process, there is additional opportunity for contamination of a product. When postpasteurization blending is used, there is no further pasteurization, and any organisms mixed in from either portion remain to contaminate the consumer. The postpasteurization process works well unless that is a source of contamination for one of the portions to be blended. Apparently, in the large outbreak discussed here, contamination occurred in this way, although investigators were never able to prove the absolute specific cause of the contamination of the low fat milk which was consumed to produce the infections. It was finally surmised that there was a cross-connection in the processing plant which could in some instances create a backsiphonage from a line carrying unpasteurized milk. The backsiphonage could contaminate pasteurized milk being blended to produce the low fat product. Potential problem areas in the producing plant were modified so that deficiencies were corrected. Some dairy plants still allow the use of postpasteurization blending to produce low fat milk products. For this to occur, the FDA must approve the

FIGURE 1.2   Dry foods which have been processed and packaged may have been contaminated in process by handling. Some (yeast) will be cooked before consumption. Others (milk and pepper) may or may not be cooked before consumption.

proper safety procedures for use. In the case of this outbreak, it was not a result of the failure of the pasteurization process. It was rather a failure in the engineering of the machinery used to process consumable milk in large quantities and at a price acceptable to the consumer.[17,18]

Processing foods of any kind offers opportunity for the food to be contaminated, either from the environment, from a worker carrying the organism, or from a worker infected by the organism. Foods contaminated during processing are often not treated in such a way as to remove the contamination before being offered for sale to the consumer. Examples of some foods packaged in such a way that contamination may be retained are seen in Figure 1.2. Other foods which may have been contaminated during processing, as in Figure 1.3 will be cooked before consumption and are less dangerous than those shown in Figure 1.2, which may not be cooked again to remove the contamination. Care in handling of such foods should be taken. For example,

 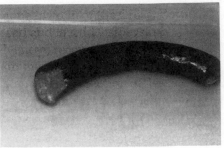

FIGURE 1.3 Processed and packaged foods may have been contaminated in process by handling or from the source. They should never be handled on a wooden cutting board.

the foods shown should never be prepared on a wooden cutting board as is seen in Figure 1.3. The most likely sources of contamination for foods during processing are tabulated in Table 1.4.

Another large outbreak of salmonellosis was reported to have resulted in Riverside, CA, in 1965 from contaminated water supplies.[19,20] That outbreak also involved approximately 16,000 persons, and although the municipal water supply was incriminated, the actual source of the contaminating organisms was never located. The contamination again was by *S. typhimurium,* and in this instance, the organism was also phage typed and determined to be type 2. Of those persons affected, at least 70 had to be hospitalized and 3 died as result of the infection. In this outbreak, there was evidence that the infectious dose of some strains of *Salmonella* may be much smaller than has been thought in the past. This outbreak was unusual in that reported waterborne outbreaks of *Salmonella* gastroenteritis have been relatively rare. Most such outbreaks which have been waterborne have involved typhoid fever rather than gastroenteritis.

The incidents described involving the spread of typhoid fever by "Typhoid Mary", and the large outbreaks of salmonellosis detected in this country are certainly illustrative of the outbreaks of diseases caused by *Salmonella* which can be expected in any developed country. Although living styles have changed, and the potential for spread of infectious diseases is now different than it was at the time of the "Typhoid Mary" incidents, the potential spread by the actions of carriers continues to be a threat to the public health in both typhoid fever and other salmonelloses. These outbreaks were investigated very thoroughly, and were likely reported fairly accurately, because of either the danger of the typhoid spread or because of the accompanying publicity surrounding the outbreak from milk. On the other hand, outbreaks of less magnitude, or in less publicized circumstances, are not as well publicized or investigated, and we therefore must rely largely on estimates to assess the danger of salmonellosis to the population in general. In 1969, the Expert

TABLE 1.4
Source of Food Contamination at Different Stages of Processing

Food type	Food state Raw	Heated	smoked	Dried	Ready-to-eat
Meats	A,F,W	C,M,T,	C,E,W,T	C,M,T,W	C,E,T,M, W
Poultry	A,F,W	C,M,T, W	C,E,W,T	A,C,T,W	C,M,T,W
Fish	A,E,W	C,T,W	C,E,T,W	T,M,W	C,M,T,W
Shellfish	A,E	C,T,W	E,C,T,W		A,C,E,M, W
Eggs	A,E,F	A,C,M, T,W		C,M,T,W	A,E,C,T, M,W
Milk	A,F,W	C,M,T, W		A,C,E,M, T,W	A,C,E,M, T,W
Cheese	A,F,W	C,E,M, T,W			A,C,E,M, T,W
Ice cream	A,E,M, W	C,E,M, T,W			A,C,E,M, T,W
Vegetables	E,F,W	C,E,M, T,W		E,C,M,W	E,F,M,T, W
Fruit	E,F,W	C,E,M, T,W		E,C,M,W	E,F,M,T, W
Coconut	E,W			E,C,M,W	E,C,M,W
Spices	C,E,W			C,E,M,W	C,E,M,W

Note: A = animal source; C = cross-contamination, handling or processing; E = environmental source; F = feed, fertilizer source; M = mishandling in processing or preparation; T = time/ temperature discrepancy; and W = Worker source.

From *Procedures to Investigate Foodborne Illness,* 4th ed., International Association of Milk, Food and Environmental Sanitarians, Ames, IA, 1987.

Committee on *Salmonella* of the National Research Council[21] estimated that there occurred in the U.S. 2 million cases of salmonellosis each year. If that was the case in 1969, for which we do not have accurate numbers of cases reported, we should assume that more cases occur at the present time, since in 1971 we had 25,694 isolations of *Salmonella* from humans reported,[22] and in 1987 we had 44,609 isolations of *Salmonella* from humans.[23,24] Only 18,649 isolations were reported in 1963.[25] Based on reports of cases from 1978 to 1982, Todd[7] estimated 2,960,000 cases of salmonellosis other than typhoid in the U.S. each year. At the time those data were collected, there were 21% fewer cases reported than currently. If we now add 21% to Todd's estimate, we arrive at 3,581,600 cases of infection annually. It is likely that the larger numbers reported more recently are due to the increased population of the country and perhaps to some improvement in reporting of these infections. That such improved reporting occurred can not readily be supported. It is also generally agreed that while the incidence of typhoid fever in this country has steadily fallen, the incidence of salmonellosis otherwise has continued to rise. In some of the early reports, the numbers of cases in an outbreak of infection appeared to be relatively small. In those incidents, the small numbers were misleading because of less efficient investigations and less efficient methods of isolation and reporting.

There has been speculation that since the incidence of salmonellosis increased from 0.4/100,000 in 1942 to 9.0/100,000 in 1966, that one of the factors involved in this increase was likely to be the increase in mass production of animal and food products. Since the very early days of investigation of the incidence of salmonella in poultry and egg products, there has been speculation that such contamination was responsible for numerous outbreaks of salmonellosis. That such contamination in dried eggs shipped to allies in World War II was responsible for increases in the incidence of salmonellosis in those countries has been reported.[26]

Other, smaller outbreaks of salmonellosis occur with some regularity, some of which may be related to each other as is seen in the following incident. In 1967, four separated outbreaks of food poisoning occurred in New York. Reports were received by the New York Department of Health over a period of 4 days. Reported symptoms suggested salmonellosis as the identity of the infection.[27] This was somewhat unusual in that with the initial report of four separate outbreaks, there is frequently no attempt to connect the incidents, and therefore, the common source for the illnesses may not be discovered. As this investigation continued, additional outbreaks were recorded and investigated until it was determined that there were 14 separate outbreaks involved, all of which had resulted from the same contaminated imitation ice cream product which had been served at 14 different banquets. In all, an estimated 1790 persons became ill. The ice cream product had become contaminated from unpasteurized egg yolks which were not cooked during the manufacture of the product. Twelve different caterers were involved in serving the product, which happened to be the only common food served at all 14 events. In this investigation it was estimated that each serving of the product may have contained as many as 11,300 *Salmonella* cells, and that the manufacturer of this product may have produced as many as 18,000 servings in six production lots of the product. These servings were provided in at least six different states on numerous occasions. Although there were unconfirmed reports of gastroenteric illness from some of these areas, complete investigations could not be made of all of the possible incidents. The authors suggested that in the outbreaks studied, 50% of those eating the ice cream became ill, and calculated that if this percentage held for all 18,000 servings, then approximately 9000 cases of gastroenteritis may have occurred from this source.[27]

Much more recently, a number of smaller outbreaks involving incompletely or inadequately cooked eggs as the transmission vehicle, has led the Centers for

FIGURE 1.4   Eggs may be contaminated externally and also internally (by transovarian transmission). Cracked shells, as seen on one egg, increase the chances for contamination internally.

Disease Control (CDC) to investigate the possibility of transovarian transmission of *S. enteritidis* in chickens. Transovarian transmission of rickettsial and virus infections from the female to offspring has been known in many animals for a considerable number of years, and such transmission of *Salmonella* in duck eggs has also been recognized. The transmission of *Salmonella* to eggs from the adult chicken has long been recognized as occurring when the egg passes through the intestinal track and is contaminated by fecal material. In transovarian transmission (Figures 1.4 and 1.5), the bacterium is transmitted by the female to the germ cell of the egg before the remainder of the egg is formed, and passage through the intestinal tract is not necessary for the contamination to occur. Transovarian transmission and spread of *Salmonella* by eggs contaminated by the chicken before the egg is laid is discussed in greater detail in Chapter 7.

The largest outbreak of foodborne salmonellosis reported to CDC before the milkborne outbreak of 1985, occurred in September 1974.[28] This outbreak was finally determined to have been transmitted by contaminated potato salad which was served to approximately 11,000 persons at a free barbecue on the first day of a fair on an indian reservation. From this contaminated food, it was estimated that 3400 cases of salmonellosis occurred. Although direct examination of the foods served was not possible while the food was fit for examination, the epidemiologic investigation established rather definitely that potato salad was the vehicle of transmission. In food preparation for the barbecue, cooked potatoes and eggs were processed and allowed to stay at ambient temperatures suitable for growth of salmonella for 13 to 16 h. Any food prepared in advance and held at such temperatures (Figure 1.6) may allow growth of salmonella after being contaminated. During processing of these

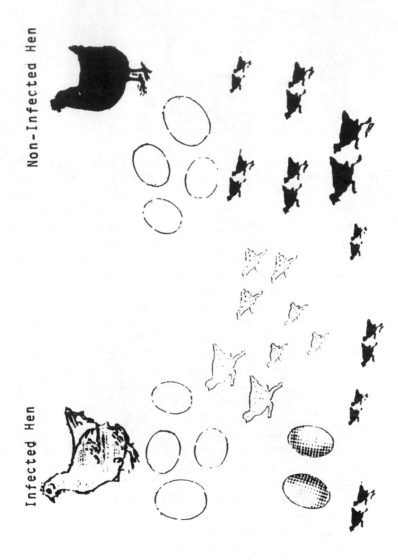

FIGURE 1.5   The infected hen at left may produce eggs which are either infected or not, whereas the noninfected hen at right will produce only noninfected eggs. If the eggs are not infected, they are more likely to hatch, whereas the infected eggs may or may not hatch. From noninfected eggs, the chicks will not be infected. From infected eggs, chicks hatched will be infected.

FIGURE 1.6 Any food, meat or other, prepared at home or commercially and held for long periods at ambient temperature may allow sufficient growth of contaminating pathogens to cause infection in the consumer.

cooked foods, 15 persons had been involved with preparation, and none volunteered information indicating recent infection, nor did any give permission for stool culture to determine carrier state. It was determined that any bacteria present in the potatoes or eggs should have been killed by the cooking process, so it was assumed that contamination occurred later from one or more of the workers preparing the food. This outbreak illustrates so well how one person may contaminate foods which will not be further treated by heat to cause outbreaks of infections. Although the investigators looked into the possibility of direct person-to-person transmission of the infection as well as transmission by the contaminated food, no evidence of direct transmission was found even in conditions where one might have expected it to occur with some frequency. This outbreak also illustrates that adequate cooking of foods will most probably eliminate the possible spread of the organisms and occurrence of infectious disease. Therefore, those steps in processes which come at the end of the procedure when no further cooking will occur are the ones which must be monitored most closely to help control salmonellosis.

The incidence of waterborne salmonellosis has appeared to increase in the decades since the 1940s. This apparent increase has been in part due to outbreaks of waterborne salmonellosis which occurred in this country in the 1965 Riverside, CA outbreak, and again in a 1976 incident in New York. The

TABLE 1.5
Steps Usually Involved in Treatment of
Potable Water

1. Presedimentation
2. Pretreatment for taste and odor control
3. Coagulation and flocculation
4. Filtration
5. Chlorination

From Guthrie, R.K., *Food Sanitation*, 3rd. ed., Van
Nostrand Reinhold, New York, 1988.

1976 incident occurred as a result of contamination of the water supply in
Suffolk, NY which accounted for 750 infections.[19] In both cases, the organism
incriminated was *S. typhimurium* as was the milkborne outbreak of infection
reported from the Midwest. Although these recent reports make it appear that
the incidence of waterborne salmonellosis is increasing, it is most likely that
this increase is more apparent than real, due to improvement in reporting and
in recognition of salmonellosis. Because of water treatment such as is described
in Table 1.5, the incidence of water-associated illness resulting from *Salmonella*
or other infections has also changed over the years.

Salmonellosis caused by contaminated shellfish has been reduced by
restricting harvesting in areas which are contaminated by sewage and by
treatment of sewage. From 1900 to 1955, there were reported 3266 cases of
typhoid fever as a result of consumption of contaminated shellfish, primarily
raw oysters (Figure 1.7). Since that time, no cases of typhoid fever from this
source have been reported, although there have been reports of 300 cases of
acute gastroenteritis infection caused by *Shigella, Vibrio, Escherichia,
Salmonella,* and other infectious agents which have resulted from consumption
of contaminated shellfish. In those states where shellfish are harvested for
human consumption, the authorities have attempted to control such outbreaks
by setting standards for shellfish growing waters and for shellfish meats for
consumption. Enforcement of these standards, and establishment and
enforcement of authorized, shellfish-harvesting areas in the coastal regions
have helped to reduce and keep under control transmission of disease by these
food organisms. It is also likely that changes and improvements in the
treatment of wastewater from domestic sewage has helped to control such
infection transmission. Although most sewage treatment does not pretend to
kill or inactivate all disease agents which may be present, the use of some
chlorination and other treatment does have the net effect of lowering the
numbers of pathogenic organisms released in sewage.

The transmission of salmonella infections through contact with contaminated
water, such as in swimming, has also decreased in the years since 1900.
Incidents in which contact with contaminated water could be proven to be the

FIGURE 1.7   Shellfish harvested from sewage-polluted waters are likely to be contaminated with a number of bacterial or viral pathogens, including *Salmonella*.

cause of the infection have mostly occurred in countries other than the U.S., although such incidents have also occurred in this country. No reports of swimming-associated, salmonella infections have been made since an outbreak in 1958 in Australia. As in the case of the reduction in cases resulting from consumption of contaminated water, it is thought that the same reasons, i.e., improved water and wastewater treatment, and reduction of cases from all causes are the most important factors which could account for the lack of cases associated with water contact.

Any outbreak of illness, whether from food or from water, should be thoroughly investigated to establish the source of the pathogen and to establish the actual transmitting agent of the organism. Procedures to investigate foodborne illness have been thoroughly described by a committee of the International Association of Milk, Food, and Environmental Sanitarians.[29] A summary of these procedures is shown in Table 1.6. The procedures to investigate waterborne illness do not vary greatly from this summary and can be found in detail in a similar publication by this organization.

TABLE 1.6
Procedures to Investigate Foodborne Illness

1. Receive complaints or alerts
2. Refer complaint to correct agency
3. Get case histories and confirm diagnosis
4. Collect food samples — handle properly
5. Test food samples — identify and type pathogen
6. Develop case definition
7. Determine if outbreak occurred by making time, place, pathogen, and person associations
8. Recommend or take control actions
9. Inform public as needed
10. Analyze data collected

From *Procedures to Investigate Foodborne Illness,* 4th ed., International Association of Milk, Food and Environmental Sanitarians, Ames, IA, 1987.

REFERENCES

1. **Carey, C.,** Mary Mallon's trail of typhoid, *FDA Consumer,* U.S. Department of Health and Human Services, Washington, D.C., June 1989, 18–21.
2. **Burrows, W.,** *Burrows Textbook of Microbiology,* 22nd ed., Revised by Bob A. Freeman, W. B. Saunders, Philadelphia, 1985.
3. **Black, P. H., Kunz, L. J., and Swartz, M. N.,** Salmonellosis — a review of some unusual aspects, *N. Engl. J. Med.,* 262(16), 811–817, 864–870, 921–927, 1960.
4. **National Foundation for Infectious Diseases,** Top 10 infectious diseases, *The Double Helix,* 13(3), 4–5, 1988.
5. **National Foundation for Infectious Diseases,** NFID revises top ten priorities for research funding, *The Double Helix,* 14(3), 4, 1989.
6. **Benenson, A. S., Ed.,** *Control of Communicable Diseases in Man,* 14th ed., American Public Health Association, Washington, D.C., 1985.
7. **Todd, E. C. D.,** Preliminary estimates of costs of foodborne disease in the United States, *J. Food Prot.,* 52(8), 595–601, 1989b.
8. **Joint FAO/WHO Expert Committee,** Microbiological Aspects of Food Hygiene, Tech. Rep. Ser. No. 598, World Health Organization, Geneva, 1976.
9. **Joint FAO/WHO Expert Committee on Food Saftey,** The Role of Food Safety in Health and Development, Tech. Rep. Ser. No. 705, World Health Organization, Geneva, 1976.
10. **Todd, E. C. D.,** Preliminary estimates of costs of foodborne disease in Canada and costs to reduce salmonellosis, *J. Food Prot.,* 52(8), 586–594, 1989a.
11. **Levy, B. S. and McIntire, W.,** The economic impact of a food-borne salmonellosis outbreak, *J. Am. Med. Assoc.,* 230(9), 1281–1282, 1974.
12. **Blumenthal, D.,** *Salmonella enteritidis.* From the chicken to the egg, *FDA Consumer,* U.S Department of Health and Human Services, Washington, D.C., April 1990, 7–10.
13. **Sanders, E., Brachman, P. S., Friedman, E. A., Godsby, J., and McCall, C. E.,** Salmonellosis in the United States. Results of nationwide surveillance, *Am. J. Epidemiol.,* 81(3), 370–384, 1965.

14. **Centers for Disease Control,** Annual Summary 1984, Morbidity and Mortality Weekly Report, Centers for Disease Control, Public Health Service, U.S. Department of Health and Human Services, Atlanta, GA, 1986.

15. **Littman, A., Vaichulis, J.A., Ivy, A.C., Kaplan, R., and Baer, W.H.,** The chronic typhoid carrier. I. The natural course of the carrier state, *Am. J. Public Health,* 38, 1675–1679, 1948.

16. **Rosenkrantz, B. G., Advisory Ed.,** *The Carrier State,* Arno Press, New York, 1977.

17. **Lecos, C. W.,** Of microbes and milk: probing America's worst *Salmonella* outbreak, *FDA Consumer,* U.S. Department of Health and Human Services, Washington, D.C., February 1986, 18–21.

18. **Lecos, C. W.,** Food poisonings from tainted dairy products, *FDA Consumer,* U.S. Department of Health and Human Services, Washington, D.C., April 1986a, 16–17.

19. **Craun, G. F., Ed.,** *Waterborne Diseases in the United States,* CRC Press, Boca Raton, FL, 1986.

20. Collaborative Report, A waterborne epidemic of salmonellosis in Riverside, California, 1965. Epidemiologic aspects, *Am. J. Epidemiol.,* 93(1), 33–48, 1971.

21. Committee on *Salmonella,* An Evaluation of the Salmonella Problem, National Research Council, National Academy of Sciences, Washington, D.C., 1969.

22. **Centers for Disease Control,** *Salmonella* Surveillance, Annual Summary, Centers for Disease Control U.S. Department of Health and Human Services, Atlanta, GA, 79-8219, 1977.

23. **Centers for Disease Control,** Summary of Notifiable Diseases United States 1986, Morbidity and Mortality Weekly Report, Centers for Desease Control, U.S. Department of Health and Human Services, Atlanta, GA, 35(55), 1987.

24. **Centers for Disease Control,** *Salmonella* Surveillance, Annual Summary, Centers for Disease Control, U.S. Department of Health and Human Services, Atlanta, GA, 1987.

25. **Aserkoff, B., Schroeder, S. A., and Brachman. P. S.,** Salmonellosis in the United States — a five-year review, *Am. J. Epidemiol.,* 92(1), 13–24, 1970.

26. **Dack, G. M., Ed.,** *Food Poisoning,* University of Chicago Press, Chicago, 1956.

27. **Armstrong, R. W., Fodor, T., Curlin, G. T., Cohen, A. B., Morris, G. K., Martin, W. T., and Feldman, J.,** Epidemic Salmonella gastroenteritis due to contaminated imitation ice cream, *Am. J. Epidemiol.,* 91, 300–307, 1970.

28. **Horwitz, M. A., Pollard, R. A., Merson, M. H., and Martin, S. M.,** A large outbreak of foodborne salmonellosis on the Navajo Nation Indian Reservation, epidemiology and secondary transmission, *Am. J. Public Health,* 67(11), 1071–1076, 1977.

29. *Procedures to Investigate Foodborne Illness,* 4th ed., International Association of Milk, Food and Environmental Sanitarians, Ames, IA, 1987.

30. **Guthrie, R.K.,** *Food Sanitation,* 3rd. ed., Van Nostrand Reinhold, New York, 1988.

Chapter 2

TAXONOMY AND GROUPING OF THE *SALMONELLA*

The 8th edition of *Bergey's Manual of Determinative Bacteriology*[1] includes the *Salmonella* in the Family Enterobacteriaccae, a practice which was continued in the newer *Bergey's Manual of Systematic Bacteriology*, Volume 1.[2] It was then recognized that many workers used the designation Coli-Typhoid Group, even though not all members of the family are pathogenic or even found in the same general habitats. This group, whether termed family or some other taxonomic designation, includes a wide variety of organisms which are ubiquitous in nature, with many having well-recognized, common locations or habitats. Included within the group are nonpathogenic bacteria, animal pathogens, plant pathogens, and opportunistic organisms. Although of greatly diverse characteristics or, in many instances, of very similar characteristics, these organisms can be separated into strains with great detail and specificity on the basis of biochemical reactions, serological analysis, and sometimes these separations can be specifically supported or even enlarged by phage typing the bacteria.

The *Salmonella* are named after D. E. Salmon, an American bacteriologist and veterinarian. Currently, these organisms are classified as Genus III in the Family Enterobacteriaceae. Some of the characteristics of the *Salmonella* which have made possible the outbreaks of disease reported at various times through history are the same as those used by taxonomists for the classification of these bacteria as members of the Enterobacteriaceae. These characteristics are not always stable in microorganisms, and therefore, nomenclature and classification of isolates which are to be included within the genus *Salmonella* remain somewhat controversial. A characteristic permitting a wide range of growth and activity in different environments is that of being a facultative anaerobe. These gram-negative, rod-shaped bacteria are included as members of the Family Enterobacteriaceae by all authorities on bacterial classification and nomenclature. In the U.S., the authority of this classification continues to be *Bergey's Manual of Systematic Bacteriology*.[2] This authority places these bacteria in a number of different species, while in other countries, a different nomenclature is used. One World Health Organization (WHO) publication in 1988 stated: "The genus *Salmonella* contains only one species."[3] In recent years, there has been more and more debate about the general taxonomic position of the *Salmonella*, in general, and of *Salmonella* species, in particular. Of the Enterobacteriaceae, the *Salmonella* and *Shigella* are almost all pathogenic, whereas the majority of strains of the other organisms are not. The *Salmonella* and *Citrobacter* are able to utilize citrate as a sole carbon source, whereas other genera require a more complex source of this nutrient, and the *Salmonella*, other than *S. typhi*, are almost always aerogenic as concerns production of gas in the acid fermentation of carbohydrates.

It is generally agreed that, as stated by LeMinor and Rohde,[4] "scientifically, none of the present methods of nomenclature of *Salmonella* is satisfactory". The International Enterobacteriaceae Subcommittee has still not given clear guidance for the naming of different types of *Salmonella*.[2] The WHO Expert Committee[3] considers that the taxonomy of this group is now established on a scientific basis with the proposal of LeMinor et al.[5] and states without qualification that the genus *Salmonella* contains only one species.[3] In essence, the collaborators writing for the most recent *Bergey's Manual of Systematic Bacteriology*[2] agree when they state that the "use of 'species' names for *Salmonella* serovars is extremely useful in many fields", and continue, "As long as these serovar names are not taxonomically equated with species, this practice should be encouraged." In common practice, the isolates of *Salmonella* are almost always assigned specific names based on the Kauffman-White scheme of serological identification of these strains, without the use of the specific Kauffman-White numbers for OH and flagellar antigens. Initially, the numbers of serotypes or serovars of *Salmonella* were relatively small, numbering in the hundreds. With continued use, this system has now resulted in over 2200 serovars based on the 67 known O antigens and the numerous H antigens which are now recognized, and the numbers continue to grow.

Gram-negative cells, because of the chemical structure of the cell wall, are more readily destained after application of gentian violet dye under proper conditions. Such destained cells are then counterstained with a contrasting color of dye. The gram-staining procedure divides bacteria into two large groups and is important in beginning identification procedures. The Gram reaction of bacteria is also related to certain other characteristics of these cells,[6,7] including the *Salmonella* which are resistant to several environmental parameters. Unrelated characteristics of the *Salmonella* include growth at temperatures between 8 and 45°C in the pH range of 4 to 9. These bacteria are able to grow only at water activities above 0.94. *Salmonella* are sensitive to heat, and, generally speaking, the organisms are killed at temperatures of 70°C or above. Because of this characteristic, ordinary cooking is sufficient to destroy *Salmonella* cells if applied for times sufficiently long enough to reach this temperature throughout the food being cooked. *Salmonella* are also susceptible to the heat applied in pasteurization of milk at 71.2°C for 15 s and are resistant to drying (may survive in dust for long periods of time, even for years). The addition of salt to foods has been used as a method of preservation throughout much of recorded history, but little effect of this material is observed on the *Salmonella* since some of them have been observed to survive as much as a 20% salt environment for weeks.[3]

The Enterobacteriaceae were so named by Rahn in 1937 and are now described as intestinal bacteria. These bacteria are motile by peritrichous flagella, do not form endospores or microcysts, and are not acid-fast. A typical cell of this type is depicted in Figure 2.1. Most members of the family are easily grown on ordinary culture media, such as peptone, meat extract, etc.,

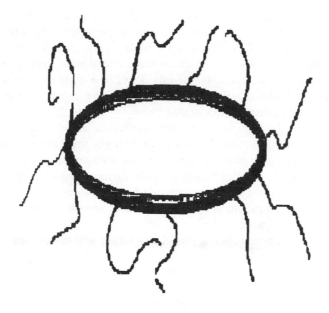

FIGURE 2.1   A typical peritrichous bacterial cell. Drawing courtesy
of Garth Morgan.

and some will grow well with D-glucose as the sole carbon source. A few may
require special vitamins or amino acids for growth. Useful characteristics for
recognition of specific members of the family are the production of acid and
gas during the fermentation of glucose and/or other carbohydrates. Enterobacteria
are found on the external surfaces and in the internal cavities of many animals
from insects to man.

Enterobacteriaceae are different from some of the other gram-negative
organisms in that as facultative cells they are able to survive and thrive in a
wide variety of environmental conditions and on a wide variety of nutritional
substrates. A facultative anaerobic bacterium can live normally in an atmosphere
of some oxygen and may prefer this state; however, if all oxygen is removed
from the environment, the organisms can survive by carrying on anaerobic
metabolism. Such a characteristic is generally descriptive of the majority of
bacteria living in the intestinal tract. This ability to survive extremes of
environmental conditions is one of the characteristics of the *Salmonella* which
make control of these organisms so difficult. Enterobacteria are found worldwide
in soil, water, and on the surfaces of fruits, vegetables, and grains serving as
food for man. Because these organisms live in such close association with the
human, they have been intensely studied to determine their medical and economic
importance. The *Salmonella* are of medical importance due to diseases produced
in the human, and in other animals. Salmonellosis is a problem not only
because it is a disease of the human but also because it is a problem in the

poultry industry worldwide. The existence of salmonellosis as an infection of poultry is important due to the economics within the poultry industry, and because contaminated poultry serves as a vehicle for the transmission of the disease to the human. In addition to salmonellosis infections in many different animal species, other bacteria in this family also cause diseases in man and other animals. *Escherichia coli, Citrobacter freundii, Yersinia* species, and *Shigella* species are commonly recognized pathogens within this family. It has been estimated that bacteria in the Enterobacteriaceae cause approximately 50% of all nosocomial infections in the human. Many of these are not actually intestinal infections, and most of them are not transmitted as foodborne infections. Salmonellosis is often considered to be the most frequently occurring infectious disease resulting from contaminated food consumption.[8] According to some authorities, all members of the *Salmonella* are pathogenic to some extent.[6] This does not mean that any time one of these organisms enters the human body in any numbers than an infectious process will start. It does mean that these bacteria must always be considered to have the potential to begin the infectious process when present in the human body.

All strains of *Salmonella* are known to infect man and many other animal species. For many years, the *Salmonella* were the only group then classified among the Enterobacteriaceae which were considered to be pathogenic. Now it is recognized that many strains of *E. coli* and *Yersinia* are consistently pathogenic and there is opportunistic pathogenicity in strains of *Klebsiella*, *Enterobacter*, *Proteus*, *Providencia*, and in some strains of *Serratia marcescens*. This latter group are among the most frequent causes of nosocomial (hospital-acquired as mentioned earlier) infections recognized in recent years.

Among the Enterobacteriaceae, the members of the *Salmonella* and *Shigella* groups are considered to be foodborne pathogens, and are most often incriminated as causes of intestinal infections. The *Shigella* are frequently associated with large-scale outbreaks of diarrheal disease in times of war, when there is a general lack of sanitary facilities and cleanliness. Other, smaller outbreaks do occur with some frequency, although not as often as do outbreaks of salmonellosis as may be deduced from the fact that in 1986, there were almost 50,000 isolations of *Salmonella* from infections reported in the U.S., while there were only some 17,000 isolations of *Shigella* from infections reported.[9] These figures, as in the case of many figures involving reported infections are not likely to be absolutely accurate, because neither disease in the adult may be very severe, and therefore, may not be diagnosed or reported in many cases.

The basis of biological classification, and including microbial classification, is the species, which gives the student or investigator a name and an identity to work with in studying and in communication about the organisms. A species, biologically, is a group of organisms capable of interbreeding. In microbiology, in the past, this definition has not always had significant meaning because these cells were not thought to have sexual reproductive pathways for which interbreeding was required for cell multiplication. With the knowledge that some bacteria at least, did have a sexual reproduction expression, the original species concept began to apply more directly to these organisms. As more has been learned about these organisms, it is accepted that many bacteria have mating types and sexual reproductive pathways. The idea of interbreeding, however, is still difficult to apply to the bacteria and other microorganisms. Other characteristics need, therefore, to be delineated for establishment of species.

For convenience, the concept of families of bacteria have been used to group bacteria according to specific characteristic and ecological considerations. The ecological considerations have not always seemed to fit well since some organisms appear to be identical except for the production of specific enzymes to attack certain substrates in their environment. In the Enterobacteriaceae, the species have generally included a group of organisms with related, but not identical, characteristics which cover a wide range of variation. These organisms appear to undergo several types of genetic recombination, and therefore, many intermediate strains are produced. In fairly recent times, a system which was actually begun in the 18th century has been revived, using a system of characters which have been determined for each organism and which can be grouped to allow formation of classes of organisms, without giving more weight to one character than to another. This system is termed numerical taxonomy and is probably not used extensively because its use is laborious, requiring special techniques and training. Also used in some laboratories, where the capability exists, is the use of the determination of DNA homology to group organisms. Again, the system is not used extensively because of the special techniques, training, and capabilities required.

The simplest form of identification and nomenclature which can be used in bacteriology has been based on fermentation and other metabolic reactions following staining and morphological characteristics. The latter may often be omitted in identification because certain growth and metabolic reactions are identified with some of the group staining reactions. With added knowledge of plasmid function, it is now recognized that these metabolic functions are also frequently based on the genetic characteristics of the bacteria. These reactions, together with some serological reactions, have been used in the past for establishment of species in the *Salmonella* resulting in much controversy and confusion. Even with this use, final identification often depended on serological reactions to detect the antigenic composition of the bacterial cell. Antigenic mosaics have been most completely documented for the organisms identified as *Salmonella*.

The WHO Collaborating Center for Reference and Research on *Salmonella*[5] has proposed as a basis for the taxonomy and corresponding nomenclature of this genus, the primary use of the Kauffman-White scheme which the Enterobacteriaceae subcommittee considers to be overidingly important, and states that, in general, new descriptions of subgroups should be designated by those formulae.[2]

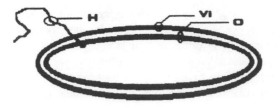

FIGURE 2.2   Location of salmonella antigens. Drawing
courtesy of Garth Morgan.

The Kauffman-White scheme is based on antigens — molecules, usually
protein, which when introduced into the tissues of a competent animal will
stimulate the production of a specific response (either antibody molecules
specific for the antigen, or specifically reactive cells). Each antigen molecule
contains a certain number of chemical groupings (determinant groups) which
ascribe immunological specificity to the molecule because of the peculiar
atomic and molecular arrangements of the chemical structure. The
immunological specificity of the molecule is easily modified by changing a
relatively small part of the large protein molecule. The specificity of the
antigen is determined by the chemical make up and the arrangement of the
structure of this portion of the molecule (the determinant group). Each microbial
cell is composed of many antigenic molecules which comprise its structure,
as each antigen molecule is comprised of many determinant groups. Each of
these then forms a mosaic, the antigens forming the antigenic mosaic of the
bacterial cell. Each antigen and each determinant group of the antigen stimulates
the production of the specific immune response (in the case of the Kauffman-
White scheme, antibodies), and the resulting immunological reaction is the
production of many antibody molecules, some of which are of different
specificity. In the bacterial cells used, antigens are generally found in the
structure of different parts of the cell (Figure 2.2). Cell wall or intracellular
antigens are termed somatic antigens, and form the structure for the cell. In
the *Salmonella*, these antigen specificities and the occurrence of certain
specificities determine the group assignments of the organisms. For example,
Group A contains somatic antigen (O antigen) 2, Group B contains somatic
antigen 4, Group C1 contains somatic antigen 7, Group C2 contains somatic
antigen 8. This is not to say that these cells do not contain other antigens. In
addition, all strains in Group B contain somatic antigen 12, as do the strains
in Group D, and Groups C1 and C2 organisms all contain somatic antigen 6,
as well as others mentioned above. Other antigens of different specificity are
a part of the structure of both the cell wall and the flagella (H antigens) of
motile organisms, and in addition, those bacterial cells which possess capsules
or envelopes will be found to possess a third variety of antigen (the Vi or K
antigen). The Kauffman-White scheme arranges the *Salmonella* into serovars

or types based on antigens present in the envelope, in the cell wall, and in the flagella. Although many strains or serovars of the *Salmonella* may contain common O or K antigens, the pathogenic characteristics of the bacteria may be very different. For example, *S. typhi* contains antigens in common with all organisms in Groups A, B, and D, but this strain is much more rigidly adapted to the human host and almost invariably produces an enteric fever type of infection more than the other serovars do. The occurrence of the Vi antigen in *S. typhi* sometimes masks reaction of other O antigens which are present, and the presence of this antigen is thought to be much more closely related to pathogenicity than are the others.

In the *Salmonella* and other enteric bacteria, the somatic antigens are comprised of lipid-polysaccharide-polypeptide complexes which make up the endotoxins found in these organisms. These are termed the O antigens, and specificity is contained in the polysaccharide portion of the molecule. Antigens located in the flagella are termed the H antigens, and those in the envelope in the *Salmonella* are termed the Vi antigens (Figure 2.2).

With current techniques, antibody preparations of extreme specificity can be produced for use in establishing the exact identity of any isolate recognized. By use of this scheme, currently there are over 2000 serovars recognized in the genus *Salmonella*. In addition to the O and H antigen components of most *Salmonella* strains, there is an additional antigenic component present in a few other strains, notably *S. typhi*, *S. paratyphi* A, and *S. paratyphi* C. This antigen component is termed the Vi and is similar in composition to the O antigen. Although similar, the Vi antigen differs in being more susceptible to heat. It is generally assumed to be the "virulence" antigen because its presence is believed to be indicative of the virulence of the organism, and specific antibody to this antigen is thought to protect against an organism producing it.[6] Although the reagents and techniques for specific identification of these serovars are available, there are many cases when the full potential of the scheme has not been utilized, but rather, the isolates were generally classified into the group *Salmonella*, and a species was established on the basis of biochemical reactions. The use of the Kauffman-White scheme will divide a group of organisms which would otherwise be considered a homogeneous species into different serovars. The O antigen is used for grouping organisms in the serological identification schemes, with the formation of 13 groups, some of which have common antigens, all of which have identifying O antigens (Table 2.1), and the last of which is made up of those organisms which do not fit elsewhere. About 98% of all strains fall into the first eight of these groups. The O antigens are designated by Arabic numerals, and H antigens by lower case letters in Phase 1 and Arabic numerals in Phase 2. The antigenic make up of an organism is worked out by determining the components of the antigenic mosaics present on the cell and on the flagella. These are determined by serological testing combined with reciprocal absorption to remove extraneous antibodies from the serum (Table 2.2). Each of the serological groups may be defined in antigenic

TABLE 2.1
O Antigen Serogroups in the *Salmonella*

Group A	O Antigens, 1, 2, 12
Group B	O Antigens, (1), 4, (10), 12
Group C1	O Antigens, 6, 7
Group C2	O Antigens, 6, 8
Group D	O Antigens, (1), 9, 7
Group E1	O Antigens, 3, 10
Group E2	O Antigens, 3, 10
Group E3	O Antigens, 3, 19
Group F	O Antigens, 11
Group G	O Antigens, (1), 13, 23
Group H	O Antigens, (1), 6, 14, 15
Group I	O Antigens, 16
Others	

Note: Antigens listed in parenthesis occur in some but
not all serotypes of the group.

TABLE 2.2
Absorption of Antiserum for Serological Use

	Reaction with bacteria	
Serum specific for	**A**	**B**
Bacterium A	4+	1+
Bacterium A, absorbed with B cells	3+	0
Bacterium B	1+	4+
Bacterium B, absorbed with A cells	0	3+

Note: Additional antigens, both O and H, can be used in similar schemes
for preparation of specific sera for bacteria containing more antigens
in the mosaic.

terms by listing the antigenic formula determined for the mosaic. For example, one strain of *S. enteritidis*, serovar Kiel, is designated as Serogroup A:1,2,12;g,p (1,2,12 are O antigens; g, p, are H or flagellar antigens). Another strain of *S. enteritidis*, serovar Nitra, is designated as Serogroup A:2,12;g,m. When more precise epidemiologic investigations are needed to determine sources and pathways of the spread of these organisms in disease outbreaks, then it becomes necessary to follow this Kauffman-White scheme and accurately identify the particular strain of bacterium which is implicated in the outbreak, as well as to establish the source of the organism. Such identification is accomplished by comparison to establish antigenic formulas, some examples of which are shown in Table 2.3. Such identifications have been established for the large

TABLE 2.3
Some Representative Antigenic Formulas

O Group	Former species name	Antigenic formula
D	S. typhi	9,12,(Vi);d; _
A	S. paratyphi A	1,2,7;a; _
B	S. paratyphi B	1,4,5,7;b;1,2
C1	S. paratyphi C	6,7,Vi;c;1,5
C1	S. choleraesuis	6,7;c;1,5
B	S. typhimurium	1,4,5,12;i;1,2
D	S. enteritidis	1,9,12; g,m; _

outbreaks of the infection which have been reported in recent years and in many of the smaller ones as well. When the identifications have been established, the serotypes are then referred to by species names as a matter of convenience, and more recently, it is often recommended that we consider these to be serovar names only and not elevate them to species status. Certain antigenic combinations have been produced in the laboratory in some instances, and it is likely that these occur naturally as a result of conjugation, transduction, mutation, or loss variations. Following phage invasion, lysogenicity, in some cases, can produce changes in the O antigen characterization of some strains, but the use of formulas for nomenclature does not require that changes in antigenicity confer different species or subspecies names on the organisms in question. The specification of O antigens of the *Salmonella* are determined by the composition and structure of the polysaccharides which constitute a part of the structure of the cell surface. Such polysaccharides are also modified during smooth to rough changes in isolates, as well as during mutations and bacteriophage conversions.

The *Salmonella* are fairly homogeneous in that they resemble each other in most characteristics more than they resemble other organisms which may be classified as enteric bacteria. One characteristic, the antigenic composition of different strains, which is most useful in identification and recognition of the organisms is known in much detail, and it has been possible to identify three different types of antigenic variation in these molecules. Of these, one is a completely reversible variation and is seen in the H antigens in that when a culture is plated out and slide agglutination is done on individual colonies, one colony may be agglutinated by one H antiserum, and another colony be agglutinated by another of completely different specificity. It is apparent that each bacterium does not contain both H antigen types, but one cell contains one type, and the second another. When a colony is picked to liquid medium, after very few transfers, it is found that the culture has reverted to a 50:50 ratio of the two types of H antigen. This type of immunologic variation is called phase variation, and the antigenic types are transitory in that they may revert to a biphasic state. Variation toward a monophasic strain has also been observed;

however, the biphasic state is more common. In phase variation of H antigens, it has been observed that Phase 1 is specific in relatively few types, while Phase 2 is nonspecific and occurs in many types. Phase variation is somewhat more complex than the description given above involving three types of phase variation which are, however, not of concern in the practice of control of *Salmonella* infections. In the O antigen complex of the group, the same kind of phase variation has not been observed; however, there is some variation in that O antigens of certain types (i.e., 12 and 6, are expressed in two or three ways), and these expressions are observed to exhibit phase variation. Regarding the Vi antigen, this molecule is most often present when the organisms are freshly isolated and is frequently lost within a few transfers in artificial culture. It is believed that such loss is a form of phase variation comparable to those observed with the H and O antigens. The presence of large amounts of Vi antigen in those strains which possess it has been observed to completely block agglutination of cells in O antiserum; the O antigens in such cases apparently being completely covered and masked.[2,6]

Antigenic variation in the H antigens can be artificially induced if the organisms are cultured on semisolid medium containing certain specific antisera. In this case, the antigen to which the antiserum is specific will be lost from the culture. Biphasic organisms can be induced to become monophasic by the same culture process in the presence of specific antiserum.[6]

As with many other types of bacteria, the *Salmonella* are observed to undergo smooth-rough (S-R) dissociation. Dissociation involves a modification of the colonial morphology and a loss of virulence. The dissociation is observed to involve loss of specificity of the O antigens on the surface of the cell, although the H antigens apparently remain unchanged.[6]

When the presence of a lactose nonfermenting, gram-negative bacillus has been detected in a food product or in a clinical specimen, the exact identification may very well involve a great amount of detailed testing requiring considerable expertise in several techniques. The organisms may first be run through a battery of metabolic tests to attempt to establish some degree of identity. Following this, most laboratories can then handle the serological testing to determine the O group of the organisms, and perhaps will have on hand more specific antisera to attempt to determine a more exact type for these organisms. The ability to test for all the possible variations of the H antigens is the step at which many laboratories will be stymied because of the variation between specific and nonspecific phases of the H antigens. The terms specific and nonspecific refer to the transitory immunological types of antigen in the flagella termed the specific phase as those characterized by the presence of specific flagellar antigens, and nonspecific, as those characterized by the presence of nonspecific antigens. Phase variation is somewhat more complex than this, in that three types of variation have been characterized, and there appears to be another type of variation which has not yet been named.[6] Most often, many of the initial serological tests will be performed as slide agglutination tests, and for confirmation and more positive determination, the tube agglutination tests are required. The *Difco Manual*, 10th ed.,[10] and the *Bergey's Manual of Determinative Bacteriology*, 8th ed.,[1] contain excellent, complete tables of serological types based on both O and H antigen determinations, as well as identification of the Vi in those strains containing it. Tables from the *Difco Manual*[10] are repeated in the Appendix B of this book.

Although serological identification of *Salmonella* isolates is most often used, and in the majority of clinical cases is sufficient for the identification of isolates, an additional identifying or typing method is required when origin and characteristics of an outbreak of infection must be determined. This method involves the use of bacteriophages or bacterial viruses. Illustration of bacterial infection by phage is shown in Figure 2.3. Phage typing is currently being used, in some instances, in the investigation of the presence of contamination by *S. enteritidis* in eggs and poultry flocks in the U.S. A program of the U.S. Department of Agriculture (USDA) is attempting to control the spread of this contamination and is attempting to prevent the entrance of Phage type 4 from Europe. It appears that all bacterial strains are susceptible to infection by specific bacteriophages or bacterial viruses. Specificity of this infection is as extreme as that involved in serological reactions, and when needed to determine the origin and the characteristics of spread of an outbreak of infection, some serovars of *Salmonella* can be phage typed to establish identity. Although most serovars, as evidenced by the fact that transduction can occur so broadly, are susceptible to infection by phages, only a certain limited number have been successfully typed by this method because of the need for specially trained personnel and the high cost of these typing procedures. In the case of *S. typhi*, 33 specific phage types were recognized in the past and used in typing strains of this species. More than 30 types and subtypes of *S. paratyphi* B phages are available. All of these phage-typing schemes are specific for the Vi antigens present on these organisms. Such specificity indicates the large numbers of reactions possible. The lack of the Vi antigen on many strains apparently restricts the number of phage types known at any time. Phage typing of strains will ultimately lead to an even greater number of *Salmonella* strains than are now recognized on the basis of serological reactions, if results with these two species are an indication. Serovars of *Salmonella* which can be divided into additional types by the use of phage typing include typhi, paratyphi B, enteritidis, typhimurium, and panama. More phage-typing systems are regularly being developed (including schemes for the serovars which are frequently placed in different groups by different investigators, such as enteritidis, dublin, hirschfeldii, and schottmuelleri). Others are likely to be added at fairly frequent intervals depending upon the importance of additional epidemiological evidence in studies of outbreaks.[3] This typing is most often used in epidemiological investigations, particularly those involving a carrier for one of the *salmonellas*,[6] and particularly when the organism in question is *S. typhi*, or one of the paratyphoid strains. As noted in Chapter 7,

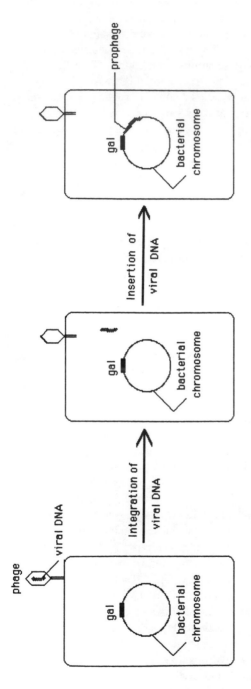

FIGURE 2.3   Bacteriophage infection of bacterial cell. Drawing courtesy of Garth Morgan.

it is a critical part of the program being attempted to limit the spread of egg-transmitted enteritidis infections in this country. Phage typing is not a process which can be carried out by all laboratories and particularly not by those dealing with the *Salmonellas* either from clinical specimens or from foods. It is rather a very complex series of reactions which are generally carried out only by *Salmonella* centers, and the International Salmonella Centre at the Pasteur Institute in Paris works with these in a consultative way.[11] Attempts to classify strains by phage typing done by inexpert personnel may lead to false and misleading grouping or classification and will serve to add confusion to an already confusing field of study.

Phage infection of the *Salmonellas* is related to a phenomenon which occurs in many strains of *Salmonella* and helps many of them to adjust to the environment — transduction. Transduction is the carrying of a fragment of a chromosome from one bacterial cell to another in the process of infection of a bacterium by a temperate bacteriophage. In this case, both strains of bacteria must be susceptible to the phage, and the phage must pick up a portion of the chromosome in the initial infection. The newly formed phage must then be capable of forming prophages in the newly infected cell, and must transfer the bacterial chromosome fragment to the newly infected cell in the process of production of lysogeny. Lysogenization by some phages may produce changes in the O antigenic mosaic of some *Salmonella*. In Groups A, B, and D, the presence of factor O1 is associated with lysogenization,[12-14] which may also be used in phage typing, but the presence or absence of this factor in these organisms does not change the name of the organism. In the case of other phages and other strains, the names of the organisms are changed.[15] Transduction may be either generalized or restricted. If generalized, the phage has an approximately equal chance of carrying any segment of the donor bacterial cell chromosome. If restricted, the transducing phage carries only those segments immediately adjacent to the site of the prophage attachment. The results of restricted transduction may be summed up in this way: "When a transducing particle is adsorbed by a recipient cell, it injects its DNA in the normal fashion. The recipient thus receives a segment of the donor's chromosome as a part of a phage genome; the latter integrates with the recipient's chromosome to become a prophage. The transduced donor genes are expressed in the recipient cell, even though they are inserted within a prophage."[16] The same sort of thing happens in generalized transduction with the exception that a part of the donor cell DNA is degraded within the recipient cell, and those portions which are left determine the transduced characters.

In early days, organisms were named by indicating the disease and the animal from which the organisms were isolated. Such a practice indicates that the organisms are most frequently found in association with that animal species. This is often not true, and the organism is found to be pathogenic for many other animals, including the human. The use of names implying animal species adaptation is very common in veterinary bacteriology, since the salmonelloses observed there are almost entirely of the zoonoses variety; that is diseases of animals transmissible to man.[17] Such names which have been assigned in the past, and which continue to be used in many cases include *S. choleraesuis* (commonly found in the pig), *S. typhimurium.* (commonly found in the mouse), *S. abortusequi* (commonly found in the horse), and *S. abortusoni* (commonly found in sheep). Perhaps the most common animal species specificity is observed in *S. typhi*, a species most frequently isolated from man, but also found in other animals as well. Although names based on animal species adaptation may often still be used, it is now recognized that most serovars, from whatever sources, are pathogenic for the human, as well as for many other animal species. In more recent practice, species names have been assigned to correspond to the geographical location of first isolation. Examples of this practice include *S. london, S. panama, S. heidelberg*, etc. The use of antigenic formula designations is gradually replacing or at least adding to these geographical usages.

It is obvious that within a group of organisms as closely related as the 2000+ serovars of salmonella, but different in at least some antigenic component, there must be great adaptability and susceptibility to genetic change. One type of genetic change which occurs in *Salmonella* cells is transduction. As described earlier, DNA is exchanged between bacterial cells by means of a temperate virus infection. In the production of new virus particles, a fragment of bacterial DNA is included in the structure of the particle. When this virus then infects a new bacterial cell, the carried bacterial DNA is deposited within the second cell and becomes a part of that cells genetic material, controlling inherited factors affected by that fragment of DNA. Some phages, or bacterial virus particles, mediate generalized transduction (that is, any bacterial genes may be transferred), while other virus particles mediate only specialized transduction (in which only certain specific bacterial genes may be transduced). In bacterial transformation, on the other hand, the DNA reaches the recipient bacterium without any carrier, whereas in transduction, the DNA is coated by the virus particle and is carried by the virus during the process of infection. Transduction is a process occurring in many more bacterial types that does transformation. In the *Salmonella* any selectable gene of the bacterium can be transduced, whereas in *E. coli*, only certain genes can be transduced.[16]

The *Salmonella* group of organisms are also known to transfer genetic material between cells by conjugation. Bacterial conjugation is the process of attachment of two different bacterial cells, with a subsequent exchange of some portion of the DNA making up the genetic material of the cells. Involved in conjugation in many bacterial strains are the genetic inclusions termed plasmids. It is generally believed that plasmids are those DNA inclusions which are readily exchanged in the process of conjugation. In recent years, the study of plasmids and their involvement in transfer of resistance to antibiotics has lead to many reports identifying mechanisms of resistance to antibiotics in microorganisms.[16] In the bacterial cells able to undergo such exchange, those cells which are able to initiate conjugation are called donors and contain a sex factor known as a pilus (plural, pili). The *Salmonella* appear to behave in conjugation in much the same way as has been so thoroughly studied in the *E. coli*. Since bacteria in the *Salmonella* group are generally pathogenic, the availability of a sexual mode of reproduction and transfer of genetic material seems to provide a broad avenue for the variation, and in essence, the creation of new strains of *Salmonella*, which may in part account for the large number of antigenic types now recognized. If those genetic characteristics which are primarily responsible for the determination of virulence were known, then this system would provide an excellent model for the study of the determination of virulence in bacteria. Unfortunately, such genetic control factors are not recognized, and we are not able to use this as a means of study. Since we can recognize only those characteristics which are readily selected in such studies, it is now impossible for us to predict new hybrid strains which may appear as potential pathogens. Studies have also shown that some genetic factors can be exchanged between certain types of *E. coli* and of *Salmonella*. Thus it appears that some pathogenic strains may be able to pass at least some genetic traits to nonpathogenic strains, thereby increasing the number of potentially pathogenic bacterial types. Pathogenic strains of this kind have not been conclusively demonstrated.

Plasmids are small, extrachromosomal, genetic elements present in bacteria with which the host cell can dispense under ordinary conditions of growth. These genetic elements are similar in many properties to bacteriophages. The primary difference is that the plasmid is not enclosed in a specific outer coat, and is apparently a native part of the bacterial cell. These structures are most significant in the *Salmonella* because they are of major importance in carrying genes for antibiotic resistance and for pathogenicity factors in these organisms. Since many of these genetic elements mediate gene transfer between bacteria, they may well be a cause of the production of new bacterial strains and are known to affect the structure of the lipopolysaccharide involved in the specificity of the somatic antigens used in the serological identification of the salmonella. This is a factor of extreme importance in the *Salmonella* group, already known to consist of more than 2000 types.

Many plasmids of gram-negative bacteria are conjugative and carry the genes which mediate their transfer in the process of conjugation. No cytoplasm is exchanged in this process, but the cell undergoes cell division, and two plasmid-containing cells exist where there was only one before. The plasmid involvement in the extrusion of the sex pilus signifies the extreme importance of plasmids in the development of new types of organisms within any strain of bacteria involving this mechanism. Plasmid genes are known to determine cell properties of drug resistance, virulence, production of antimicrobial agents, metabolic activities, and chromosome transfer. It is therefore obvious that plasmids are most important to all aspects of bacterial identification which have been used for the *Salmonella* organisms.

Bacteria, like all living cells, are subject to spontaneous mutations which involve molecular change of the genetic mechanism of the cell. Since bacteria have such a short generation time, and bacterial populations increase logarithmically when growing, the frequency of bacterial mutation may appear to be greater than that in other types of organisms. This is, however, only a matter of appearance because of the short generation time, and if properly calculated, it would be found that the mutation rates in most bacterial cultures would not differ greatly from other types of cells. That mutations occur in *Salmonella* is easily demonstrated. A mutation in one of the genes of *S. typhimurium* has been shown to cause an increase in the mutation rate for all genetic loci by a factor somewhere between 100 and 1000. Mutations can be selected for, and by devising the proper tests, can be readily recognized when they occur. Selection for mutations is a selection for the process of mutation and does not imply that a certain specific genetic change can be selectively produced at will. It is obvious in many bacteria, including the *Salmonella*, that virulence is a factor that is controlled by genetic makeup, and that as such is a factor that is subject to mutation and change on a spontaneous basis. The variation from smooth to rough in these organisms is also an example of genetic mutation which reduces the virulence of the bacterial strain being studied. In some serovars, mutation to alter the use of certain nutrients is related to virulence of the organism, and such a mutation can be recognized by observing the use of, or biosynthesis of the nutrient.[6] Other modifications of the specificity of somatic antigens may result after a mutation as reported by Kauffman.[18,19]

In bacteria, certain enzymes have been recognized and termed inducible or adaptive enzymes. Adaptive enzymes are those which are minimally produced in the absence of certain substrates. They are produced in much greater abundance in the presence of the substrate.[16] This type of enzymatic activity leads to the observation that bacteria are most adaptable organisms as concerns their environment. One of the places where such adaptability among bacteria has created the greatest problems for the human is in the case of sensitivity or resistance to certain antibiotics. Many bacteria, when forced to grow for extended periods of time in sublethal concentrations of antibiotic substances will eventually become resistant to those antibiotics, and are therefore no longer affected by larger doses of the chemical. The development of resistant strains by continual contact with sublethal concentrations of antibiotic substances is a real and constant concern of medical practice. Multiple antibiotic resistance was a characteristic of the strain of *S. typhimurium* involved in the outbreak of infection from dairy products in the Midwest in 1985.[20] More recently, further study of that strain has determined that it was resistant to tetracycline, erythromycin, clindamycin, sulfisoxazole, sulfadiazene, triple sulfa, cefoperazone, streptomycin, mezlocillin, piperacillin, carbenicillin, penicillin, ampicillin, and kanamycin. The plasmid analysis of the strain was different from other *Salmonella* strains isolated in this country in recent years, and in experimental work, transfer of plasmids from this strain to plasmid-free *E. coli* also transferred the specific resistance to the antibiotics above.[21]

Antibiotic resistance/sensitivity may also be altered by mutation in these and other bacteria. The percentage of *Salmonella* strains which were resistant to antibiotics after isolation from the human began to rise around 1960 but have since declined somewhat. It is likely that such percentages will continue to vary in response to current medical practices in combination with the current regulations of the use of animal feeds containing antibiotics. As many as 25% of *Salmonella* strains have been shown to be resistant to ampicillin, and 5% may be resistant to chloramphenicol. The use of antibiotics in *Salmonella* infections other than typhoid fever is often questioned because such use has been shown to lengthen the carrier period in recovering patients.[22]

Under the right circumstances, all *Salmonella* will cause infection in the human. Because of this pathogenic characteristic, it is essential for epidemiologic investigations that correct and repeatable typing and identification of all *Salmonella* strains be available. There has been no demonstration that any strain of these organisms has any beneficial effect on either man or other animals which they inhabit with so much frequency. Although infection may not be initiated with each contact of the organisms with man, the potential for infection is always present. Even though the organisms are ubiquitous, control is essential to avoid frequent large scale outbreaks of salmonellosis. Two most excellent examples of epidemiological investigations of salmonellosis outbreaks are the recent milkborne outbreak in the Midwest, and the waterborne outbreak in Riverside, CA in 1965. One of the best proven controls for this or any other infectious disease continues to be good sanitary practice in the production, processing, and preparation of foods for human consumption.[23] With the reduction of waterborne outbreaks of salmonellosis by the treatment of consumable water supplies, the potential for control of these organisms has been demonstrated. That reduction, however, did not eliminate, nor did it even greatly reduce, the potential for contact with *Salmonella* in the environment, as is amply demonstrated by the discussion in the previous chapter concerning the major sources of *Salmonella* contamination for the human. The human, by consumption of so many meat-containing foods, will continue to be subjected to this contact until control in food animals is accomplished. Ecologically, the organisms remain essentially uncontrolled, and until such control is accomplished, it behooves the human to exert all efforts toward assuring control through good practices of food production, processing, and preparation.

REFERENCES

1. **Buchanan, R. E. and Gibbons, N. E., Eds.,** *Bergey's Manual of Determinative Bacteriology*, 8th ed., Williams & Wilkins, Baltimore, 1974.
2. **Kreig, N. R. and Holt, J. G., Eds.,** *Bergey's Manual of Systematic Bacteriology*, Vol. 1, 1st ed., Williams & Wilkins, Baltimore, 1984.
3. **WHO Expert Committee,** *Salmonellosis Control: The Role of Animal and Product Hygiene*, Tech. Rep. Ser. No. 774, World Health Organization, Geneva, 1988.
4. **LeMinor, L. and Rohde, R.,** Genus IV. Salmonella *Lignieres 1900*, *Bergey's Manual of Determinative Bacteriology*, 8th ed., Buchanan, R. E. and Gibbons, N. E., Eds., Williams & Wilkins, Baltimore, 1974, 298.
5. **LeMinor, L., Veron, M., and Popoff, M. Y.,** Proposal for a nomenclature of *Salmonella*, *Ann. Microbiol.*, 133B, 223–243, 1982.
6. **Burrows, W.,** *Burrows Textbook of Microbiology*, 22nd ed., Revised by Bob A. Freeman, W. B. Saunders, Philadelphia, 1985.
7. **Stanier, R.Y., Duodoroff, M., and Adelberg. E. A.,** *The Microbial World,* 3rd ed., Prentice-Hall, Englewood Cliffs, NJ, 1970.
8. **Todd, E. C. D.,** Preliminary estimates of costs of foodborne disease in the United States, *J. Food Prot.*, 52(8), 595–601, 1989.
9. **Centers for Disease Control,** *Summary of Notifiable Diseases United States 1986*, Morbidity and Mortality Weekly Report, Centers for Disease Control, U.S. Department of Health and Human Services, Atlanta, GA, 35(55), 1987.
10. **Difco Laboratories,** *Difco Manual*, 10th ed., Difco Laboratories, Detroit, 1984, 772–835.
11. International Salmonella Centre, Pasteur Institute, rue du Docteur Roux 25, F-75015 Paris, France, World Health Organization, Geneva, 1988.
12. **Iseki, S. and Kashiwagi, K.,** Lysogenic conversions and transduction of genetic characters by temperate phage Iota in *Salmonella*, *Proc. Jpn. Acad.*, 33(8), 481–485, 1957.
13. **Stocker, B. A. D.,** Lysogenic conversion by the A phages of *Salmonella typhimurium*, *J. Gen. Microbiol.,* IX, 18(1), 1958.
14. **Zinder, N. D.,** Lysogenic conversion in *S. typhimurium*, *Science*, 126, 1237, 1957.
15. **LeMinor, L.,** Conversions antigeniques chez les *Salmonella*. VI. Acquisitions des facters 6, 14 par les serotypes du groupe K (O:18) sous l'effet de la lysogenisation, *Ann. Inst. Pasteur (Paris)*, 108, 805–811, 1965.
16. **Jawetz, E., Melnick, J. L., and Adelberg, E. A.,** *Review of Medical Microbiology*, 17th ed., Appleton and Lange, Los Altos, CA, 1987.
17. **Merchant, I. A. and Packer, R. A.,** *Veterinary Bacteriology and Virology*, 7th ed., Iowa State University Press, Ames, IA, 1971.
18. **Kauffman, F.,** A new antigen of *S. paratyphi* B and *S. typhimurium*, *Acta Pathol. Microbiol. Scand.*, 49, 299–304, 1956.
19. **Kauffman, F.,** *Enterobacteriaceae*, 2nd ed., Munksgaard, Copenhagen, 1969.
20. **Lecos, C. W.,** Of microbes and milk: probing America's worst *Salmonella* outbreak, *FDA Consumer*, February, 18–22, 1986.
21. **Schuman, J. D., Zotolla, E. A., and Harlander, S. K.,** Preliminary characterization of a food-borne multiple-antibiotic-resistant *Salmonella typhimurium* strain, *Appl. Environ. Microbiol.*, 55(9), 2344–2348, 1989.
22. **Benenson, A. S., Ed.,** *Control of Communicable Diseases in Man*, 14th ed., American Public Health Association, Washington, D.C., 1985.
23. **Guthrie, R. K.,** *Food Sanitation*, 2nd ed., Van Nostrand Reinhold, New York, 1988.

Chapter 3

SALMONELLOSIS — THE INFECTION

Salmonellosis in the human occurs in a wide variety of forms presenting a broad clinical spectrum. The disease may occur solely as an intestinal infection, termed salmonellosis or salmonella gastroenteritis, as a focal infection in any organ of the body, or as a systemic febrile infection. The clinical symptoms of the intestinal infection vary from asymptomatic, or no symptoms, to a most severe diarrhea with fever and nausea. Prolonged fever usually does not occur with infections other than those caused by *Salmonella typhi*, which is classified as an enteric fever. In addition to enteric fever and the more common enterocolitis (gastroenteritis), the *Salmonella* may also produce bacteremic or septicemic infections. The clinical characteristics of the different types of *Salmonella* infections are listed in Table 3.1.

Differences in the infections are largely a matter of the symptoms presented and severity of the disease. When *Salmonella* are ingested they must survive the acid pH of the stomach to set up infection. If surviving in adequate numbers, the bacteria reaching the small intestine may penetrate the mucosa of the intestine to the midlayer of this membrane where they are engulfed into the epithelial cells. *Salmonella* penetrate epithelial cells, causing an inflammatory response in the small bowel and the colon. The presence of the bacteria in this location results in an inflammatory response, and depending upon the serovar involved, the infection may progress past this tissue into the deeper layers of the mucosa of the intestinal wall. In salmonellosis, the diarrheal symptoms result from the inflammatory reaction which has been elicited in the small intestine. Some strains, at least, are capable of producing an enterotoxin which is important in the production of the diarrhea. as mentioned before.

In typhoid fever, when the organisms invade the tissues, the response elicited is monocytic in nature. The monocytes engulf the bacteria, which are not killed, but which may continue to grow within the cells. The migration of the monocytes following this growth is important in the spread of the bacteria throughout the body tissues. In the first week, Peyer's patches may form with necrosis which may cause intestinal bleeding or even bowel perforation.[1]

Typhoid fever the infection caused by *S. typhi* or, in some cases, by the *S. paratyphi* strains, A, B, or C, is a systemic infection characterized by fever, headache, enlargement of the spleen, rose spots on the abdominal surface, and constipation more often than diarrhea. Frequently, the infection is characterized by prostration and septicemia. These symptoms may be observed in other salmonelloses as well. The incubation period in such an infection may be as long as 3 weeks following ingestion of the infectious dose of organisms. Such an infection is also produced with some variation in symptoms by other

TABLE 3.1
Clinical Types of *Salmonella* Infections

Type	Incubation	Symptoms	Duration
Gastroenteritis	6–72 hours	Nausea, vomiting, fever, diarrhea	1–4 days
Enteric fever	1–3 weeks	Fever, rash, abdominal discomfort, bacteremia	1–3 weeks
Septicemia	Varies	Spiking fever	Weeks
Localized infection	Varies	Abscess	Weeks

Salmonella strains including *S. paratyphi* A (previous names have included *Bacillus paratyphosus* A and *S. paratyphi*); *S. paratyphi* B (previous names have included *Bacillus paratyphosus* B, *Bacterium paratyphosus* B, and *S. schottmulleri*); *S. paratyphi* C (previous names have included *S. hirschfeldii*); *S. barielly, S. enteritidis* variety *moscow, S. sendai, S. typhimurium, S. saint paul, S. oranienburg, S. hartford,* and *S. panama.* In some parts of the world, the enteric fevers caused by the paratyphi strains are common, and these organisms are included in the vaccine used to control typhoid infection. In this country however, such infections caused by the paratyphi strains are relatively rare, and the organisms are not included in the vaccine in use. The enteric infections caused by the last strains listed are even less frequent than those caused by the paratyphi strains. Actually, any *Salmonella* strain or serovar may cause such a systemic infection, but typhoid is considered by many to be caused only by *S. typhi*, while other enteric fever infections are considered to be typhoidal in nature, or a continued fever type of disease.[2]

The infectious nature of typhoid fever has been known for more than a century, and the causative organism was one of the earliest pathogenic bacteria to be isolated and characterized. The transmission of *S. typhi* most often occurs by food or water which has been contaminated by feces or urine from infected humans. The organism appears to be a parasite only of the human, and if multiplication occurs outside the human body any increase in numbers is negligible. The epidemiology of typhoid fever literally depends upon the transmission of fecal or urine contamination from an infected individual to something to be ingested by susceptible individuals. The control of typhoid fever was greatly enhanced by the treatment of water supplies for general consumption; however, this was not an absolute control since the organism does readily establish a carrier condition in recovering patients, and in some of these individuals the condition may be permanent. At the present time, typhoid is most often transmitted by contaminated foods, and according to many, the foods most likely to be contaminated are meats. Since the contamination for transmission must come after the food has been processed or prepared for consumption, those foods which are allowed to stand for some considerable period of time after being cooked before they are served or those foods which are not heated before service are the ones most likely to be involved in this transmission. Food may be contaminated by persons who are carriers and who use less than desirable hygienic methods in preparation or who use contaminated ingredients, including contaminated water used as an ingredient or used for cleaning utensils. When these contaminations are followed by inadequate or no cooking, and the food is then allowed to stand, multiplication of bacteria in the food will be sufficient to provide an infectious dose to the consumer. Following ingestion, after the food has passed through the acid conditions of the stomach, the organisms multiply in the small intestine and the infection is established. *S. typhi* as well as other serovars are resistant to the alkaline conditions found in the intestinal tract.

S. typhi and other serovars of the *Salmonella* are essentially spread in the environment in the same ways, with the exception that typhi is adapted to the human, and most other strains are able to infect and to be spread by other animals as well. In the cycling diagram from the World Health Organization (Figure 3.1),[3] the spread of typhoid would be limited to the cycle products eaten by man. This means that at the present time, carriers are the ultimate source of the organisms which are spread to foods, and in these contaminated foods returned to susceptible humans. As is illustrated in Figure 3.1, many other sources of infecting organisms are available for other serovars of *Salmonella*, and therefore the control of these infections is expected to be much more difficult.

Because all persons differ in susceptibility to any infectious disease, and microorganisms are likely to differ considerably coming from different environments, it is most difficult to establish a minimum infective dose for any pathogen. Some of the better work which has been done in this area was published by the Safe Drinking Water Committee of the National Academy of Sciences.[4] In that account, it was stated that Hornick and co-workers[5] had determined that when 14 persons were tested with doses of 1000 *S. typhi* cells, none became infected. But when 116 persons were tested with 10,000 cell doses, 32 became infected, and with 10 million cells, 16 of 32 were infected. Other reports have agreed that a dose of 10 million cells was necessary to cause disease in 50% of normal volunteers, and with other serovars of *Salmonella* contaminating other foods, it has been suggested that fewer than 10 viable cells can cause an infection.[6] Additional studies have reported infective doses as low as 1 to 6 cells of *S. nima* from contaminated chocolate;[6] and 60 to 65 cells of *S. eastbourne* also from chocolate[7] and between 1 and 6 cells of *S. typhimurium* from cheddar cheese were calculated as the infectious doses for six patients who became ill after consuming the cheese.[8] In earlier studies, McCullough and Eisele[9,10] reported on the pathogenicity of six separate serovars of *Salmonella* in studies of contaminated spray-dried whole eggs and found that the infectious dose varied with the strain used, but in general was found to be between 1000 and 100,000 cells. Matic et al.[11] found egg powder to be contaminated with

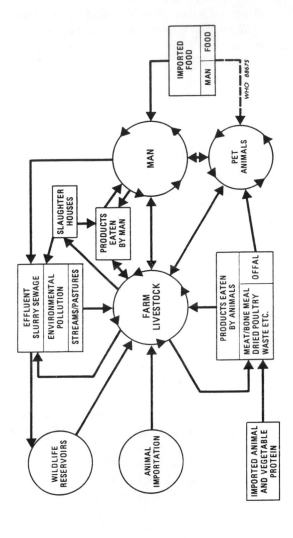

FIGURE 3.1 *Salmonella* cycling and recycling. (From Salmonellosis Control: The Role of Animal and Product Hygiene, Tech. Rep. Ser. No. 774, World Health Organization, Geneva, 1988. With permission.)

TABLE 3.2
Factors Helping to Determine Infectious Dose of *Salmonella*

Pathogen associated factors
1. Serovar; genetic composition
2. State of ingested, contaminated material (dry, moist, liquid)
3. Nature of ingested contaminated material (does it allow only survival or does it allow growth?)
4. Potential inhibitory or toxic chemicals in contaminated ingested material
5. Antibiotic status of the pathogen
6. Stress status of the pathogen

Host associated factors
1. Genetic composition of host
2. Age of host (aged and very young most susceptible)
3. Antibiotic status of host (how recently have antibiotics been administered? what antibiotics were administered?)
4. Stress status of host
5. Immune status of host

three serovars of *Salmonella*, including lille, enteritidis, and typhimurium, and found that irradiation with low dosage (1 kGy), followed by 3 weeks storage at refrigeration temperatures, was approximately equivalent to the bactericidal effect of a much stronger (3 kGy) dose of irradiation.

In a more recent review, Blaser and Newman[12] found that in 6 of 11 outbreaks when organisms were fed to volunteers, the actual infectious doses were calculated to be $<10^3$ organisms. These authors also reviewed some of the factors which may affect the infective dose of serovars of *Salmonella*, including variation within organisms. Strains which are more strongly host-adapted to other animal species, such as choleraesuis and dublin, may be more often isolated from the blood of the infected human than are other serovars. There also was some evidence found that pullorum, a serovar strongly adapted to poultry which is nonflagellated, must be administered in very high doses to result in an infection in the human. In addition to variation between pathogenic organisms, there is also suspected to be large variation in the susceptibility of the host to infection with *Salmonella*. Host factors which are important are age, other underlying illness, immune status (particularly as concerns the serovar in question), and conditions of the intestinal tract such as high acidity in the stomach of the host or the chemical nature of the vehicle in which the organism is carried.

Although it might seem desirable to establish a specific minimum infective dose for any pathogen, it is essentially an impossible task to establish one such dose for the *Salmonella* group. Indicated in Table 3.2 are the factors, involving both the pathogen and the host, which help to determine the infective dose for any individual in any one event. As is obvious, all 2000+ serovars of

Salmonella and each individual human will have a specific effect on the infective dose for each separate event.

In research with other organisms, Xu et al.[13] postulated that vibrios and coliforms may well exist in a dormant or nonculturable state in the environment, and that they cannot always be detected from that state. While this state has not yet been reported for *Salmonella*, it certainly appears likely that a similar situation can exist with these bacteria, particularly since the bacteria are known to be difficult to culture and isolate from some environments. It is apparent that all of these reports are based on results of cultures and infections which have occurred under different conditions, in persons of different susceptibility, and that the contaminants have come from different contaminated ingested materials. Differences in the materials contaminated and in the strains of bacteria which are present, as well as differences in environmental conditions all will cause variations in the infectious dose of pathogenic organisms determined at any one time.

In typhoid fever, when the susceptible human has come in contact with an infectious dose of the pathogens, there usually follows an incubation period of at least 1 week, and often of 10 days to 2 weeks. This incubation period, like the infectious dose, will vary according to conditions and strains, as well as the numbers of organisms ingested. Following the incubation period, the symptoms which appear gradually may be fever, chills, headache, myalgia, bradycardia, constipation (in about 50% of the cases), diarrhea (in about 20% of cases), and muscle soreness. These symptoms may vary in appearance and intensity with the exception of the fever which occurs in most cases. The fever usually continues to rise stepwise, reaching a maximum in 7 to 10 days, and the spleen and liver become enlarged. Peak fever levels may remain for a week or two. The white blood count is generally lower than normal, and for a short time, a rash may appear as discrete, rounded rose spots on the trunk. The rash is most likely to occur during, or shortly after the time that the organism can be found in the blood circulation. The rash persists for 2 to 5 days, and then fades, and is not often seen when the infection is caused by serovars other than *S. typhi*. In untreated typhoid fever, the mortality rate in the past was as high as 15% (and the highest rates were in those cases caused by *S. typhi*), but with antibiotic treatment this has been reduced to less than 1%. Antibiotic treatment of typhoid fever is generally required.

Typhoid is a generalized infection spreading from the entry of organisms through the gastrointestinal tract. In experimental studies, it has been found that the organisms penetrate the intestinal epithelium and move through the thoracic duct to gain entrance to the blood circulation, from where they spread to the peripheral circulation. The epithelial tissue is fairly rapidly cleared of bacteria as they move into the mesenteric lymphatics where they multiply, initiating a monocytic response. The organisms are engulfed by the monocytes, and then multiply within these cells and move into the blood stream from the lymphatics. Intestinal symptoms result from inflammatory lesions produced in the mesenteric lymph tissues (Peyer's patches). Necrosis of the intestinal tissues near these lesions may result in intestinal bleeding or even in bowel perforation. Although some of the secondary symptoms (muscle pain and continued fever) may be attributable to the endotoxin, it appears that the endotoxin has a small role in the production of intestinal pathology.[2]

Complications of typhoid infection include intestinal hemorrhage, intestinal perforation, necrosis of lymphoid tissues, hepatitis, liver necrosis, ulcerated larynx, infection of bone and joints, and inflammation of the lungs, gall bladder, peritoneum and other organs. Osteomyelitis has been observed to occur as long as 6 years after infection, indicating that the bacteria have remained viable in the tissues for these years.[1,2] This inflammation may be primarily due to the release of the endotoxic materials from the bacterial cells as these cells are killed by the body defenses. Generally, the inflammation, fever, and muscle pain and soreness will all subside as the live cells in body tissues are reduced.[14]

Typhoid fever may be diagnosed by culturing the organism from body excretions or tissues. In the first week of the infection, repeated blood cultures will frequently yield positive cultures. The bacteria do not multiply in the blood, and the situation, therefore, is not septicemia. Stool cultures may become positive by the beginning of the 2nd week. Urine cultures, if positive at all, will not be positive before about the second week of the infection when stool cultures become positive. In addition to culture, blood serum may be tested for a rising antibody titer for the pathogenic organism. A single test demonstrating the presence of antibody is not a useful diagnostic procedure, but simply indicates that the patient has come in contact with the disease organism, or has been immunized with the specific vaccine in the past. If, however, the antibody titers on subsequent tests increase, this indicates that the antigens are still present in the body, or that they have been present recently enough that specific antibody is still being produced.

The reduction of mortality rates from 15% to under 1% with treatment indicates that the infection is treatable by a number of antibiotics. This treatability, however, has not been the major factor in reduction of incidence of the disease. That factor must be recognized as the almost universal treatment of water supplies and frequent treatment of sewage with chlorination in this country. The fact that some disease remains must be attributed to carrier spread of the organism to foods. In the U.S. in 1900 there were an estimated 350,000 cases of typhoid fever, resulting in 35,379 reported deaths from this disease. In 1983, with a much larger population, there were only 507 reported cases of the disease which resulted in only three deaths. As is the case with most gram-negative, enteric bacilli, the typhoid bacillus is resistant to penicillin. The best chemotherapeutic results have been obtained with chloramphenicol, with ampicillin as the second best and the next drug of choice. Although *S. typhi* appears to be sensitive *in vitro* to a number of antibiotics, particularly the tetracycline group of drugs, it readily develops antibiotic resistance and has been observed to become resistant to both drugs when present in the bowel of patients being treated with chloramphenicol and trimethoprim-sulfamethoxazole.[2]

In typhoid fever, and some of the other salmonelloses, a complicating factor in control of the disease is the production, in some persons, of a carrier state. A carrier, in any disease, is an otherwise healthy appearing person who harbors and spreads pathogenic organisms without the appearance of symptoms. In typhoid, and some other intestinal infections, the carrier state generally involves the gall bladder as the organ in which the bacteria reside. In this location, where the bacteria are or become resistant to the bile and alkaline conditions, the organisms are somewhat protected against the defense mechanisms of the host and continue to live and multiply. In typhoid infections in this country, although the carrier condition is not considered to be common, it does occur and provides one means of maintenance of the organism in the environment in the absence of obvious cases of the disease. If a permanent condition, the organisms may not be shed constantly and, therefore, may be missed at some intervals. With immunity, the carrier condition is often permanent, unless the individual is treated to remove the pathogen. In the convalescent, the carrier condition may exist as a temporary state during the convalescent period when the person will shed the pathogen in the feces, thereby contaminating the environment. A person with poor hygienic practices may spread the organisms to others in the same environment, particularly when they work as food handlers. The most notorious case of this kind was described in Chapter 1 in the case of "Typhoid Mary" of New York.

The carrier rate is difficult to establish accurately, but in the 1940s was variously estimated to be 1:2500 to 1:3500 or higher. In 1961, the carrier rate in England was estimated to be as low as 1:100,000. In 1984, the Centers for Disease Control[15](CDC) published the carrier rate in the U.S. as being 0.03:100,000.[15] The same rate was reported by the CDC in 1980.[15] These numbers, in all cases, are likely to be lower than the actual rates since so many carriers go undetected because they are intermittent shedders of the organisms. When intermittent or light excreters of a pathogen are tested, it is most difficult to detect the carrier condition. Experts disagree as to the exact number and timing of negative cultures required to exclude the carrier state in an individual. The carrier condition has been demonstrated to be less frequent in other salmonelloses than in typhoid fever;[17,18] however, the potential for the establishment of a carrier state in any intestinal infection is always present and should be remembered in program plans for controlling such diseases. The carrier condition, frequently residing in the gallbladder or at least the biliary tract, is most difficult to cure, and antibiotics generally fail to clear it, although ampicillin has been reported to have been successful as a treatment to cure some typhoid carriers. More often, cholecystectomy (surgical removal of the gall bladder) is required in combination with antibiotic treatment.

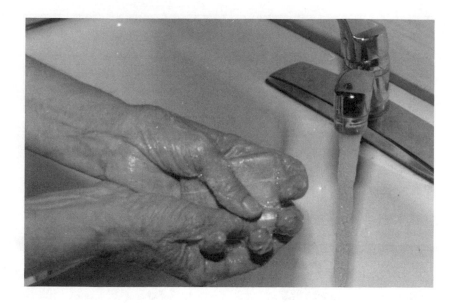

FIGURE 3.2   Frequent and thorough handwashing, particularly when handling any food during processing or preparation, is arguably the best sanitation practice available to avoid contamination of foods.

The prevention of typhoid fever is most successful with good sanitation practices: purification of water supplies; proper disposal and treatment of wastes (particularly bodily wastes from patients and carriers); good personal hygienic habits (frequent, thorough handwashing [Figure 3.2]); washing and sterilization of bed linens and clothing, particularly in patients, convalescents, medical personnel, and contacts of cases, and use of easily cleaned and disinfected utensils for handling foods as in (Figure 3.3). Patients and convalescents will shed the bacteria in feces, and frequently in urine during infection, for days to weeks after the infection subsides, providing a source of organisms which can contaminate the environment and be spread to susceptibles.

When typhoid appears in a population, it can generally be controlled by isolation and careful control of all activities and contacts of patients. Patients or convalescents should never be allowed to work in processing, preparation, or service of foods to others, and very careful personal hygienic practices should be taught to patients, contacts, and caretakers. Since the organism does not live for long periods of time in the external environment and does not parasitize other animal species, infection is most likely to be spread directly from the patient, convalescent, or the carrier to the susceptible individual. Immunization (Figure 3.4) can be used to aid in control of an outbreak of

FIGURE 3.3   Good sanitary practice involves the use of some material other than wood for cutting boards. It is most difficult to clean and disinfect wooden boards, whereas the hard surface materials are easily cleaned and disinfected.

typhoid; however, it is not very likely that an individual receiving immunization for the first time will develop sufficient immunity to ward off infection if the immunization is given after exposure to the organisms. Vaccination is no longer recommended in this country as a general preventive measure because the number of cases of typhoid fever have been so reduced (Figure 3.5), but limited specific use of immunization is of value in this country and is particularly valuable for those traveling in endemic areas, where water supplies are untreated or following natural disasters (floods, earthquakes, storms) where water supplies may have been contaminated by sewage. The vaccine used is a suspension of killed *S. typhi* cells which is given in a primary series of two injections several weeks apart. If the conditions of probable exposure remain unchanged, persons should be revaccinated after 3 years with a single booster dose of vaccine. In such cases, currently where the newer attenuated vaccine is recommended, the schedules are changed. Immunization is never an absolute preventive measure for any infectious disease, but will protect against any dose of organisms which is likely to be encountered, or at least will modify the infection to result in a less serious illness.

Several different new typhoid vaccines are being tried in attempts to find a preparation which is more efficacious, particularly in giving longer lasting immunity. One current trial is testing, on a large scale, two different kinds of

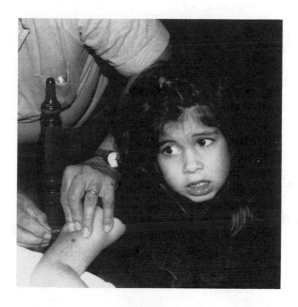

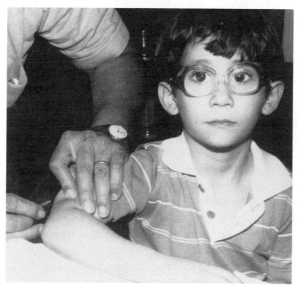

FIGURE 3.4 Immunization is one means of control for typhoid fever and is useful in endemic areas and following natural disasters which may result in contaminated water supplies.

attenuated, or live, vaccines in two locations where the disease is endemic. There are also trials of oral vaccine preparations underway in two endemic areas.[19]

**CHANGES IN NUMBER OF TYPHOID FEVER CASES IN THE
UNITED STATES FROM 1941   THROUGH 1985.
*PARATYPHOID CASES INCLUDED.**

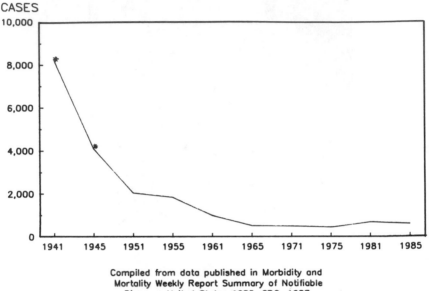

Compiled from data published in Morbidity and
Mortality Weekly Report Summary of Notifiable
Diseases United States 1985. CDC, 1987.

FIGURE 3.5   Incidence of typhoid fever in the U.S. Graph courtesy of Garth Morgan.

If the numbers in Table 3.3 are examined carefully, several things become
apparent about the infections caused by the *Salmonella*, and those other closely
related diseases which are so frequently simply classified as gastroenteritis
without benefit of diagnosis by culture. First, it is apparent that the spread of
typhoid by water and other modes of transmission has decreased since 1920,
both in total numbers of outbreaks and in numbers of cases per outbreak. One
must also assume that the numbers of deaths from any cause has decreased,
and hopefully, this may be attributed to better diagnosis and more timely and
effective treatment. A reduction in the number of cases per outbreak may
simply reflect the fact that more recently, outbreaks have been foodborne rather
than waterborne. It is not uncommon for foodborne outbreaks of disease to
show smaller numbers of cases than do waterborne outbreaks.

Unfortunately, one must also recognize that the continued use of the category
"gastroenteritis" in reporting these outbreaks and infections indicates that
specific diagnosis still leaves much to be desired. Certainly, some *Salmonella*
infections may be among the diseases included in this classification, but we
now recognize that other infections caused by *Campylobacter, Aeromonas,
Listeria, Pleisiomonas, Vibrio,* and other less readily identified pathogens may
be present in this group which are not specifically diagnosed.

TABLE 3.3
Changes in Occurrence of Salmonellosis (Including Typhoid Fever) Resulting from Waterborne Outbreaks

Years	Disease	Number of		
		Outbreaks	Cases	Deaths
1920–1925	Typhoid fever	127	7,294	435
	Gastroenteritis	11	27,756	0
1926–1930	Typhoid fever	100	3,072	234
	Gastroenteritis	17	63,902	0
1931–1935	Typhoid fever	85	2,114	140
	Gastroenteritis	25	7,664	0
1936–1940	Typhoid fever	60	1,281	80
	Gastroenteritis	91	77,403	2
1941–1945	Typhoid fever	56	1,450	46
	Salmonellosis	1	12	0
	Gastroenteritis	126	36,118	3
1946–1950	Typhoid fever	18	264	5
	Gastroenteritis	87	10,718	0
1951–1955	Typhoid fever	7	103	0
	Salmonellosis	1	2	0
	Gastroenteritis	31	5,297	0
1956–1960	Typhoid fever	13	128	3
	Salmonellosis	2	17	0
	Gastroenteritis	21	2,306	0
1961–1965	Typhoid fever	11	63	0
	Salmonellosis	2	16,425	3
	Gastroenteritis	18	20,627	0
1966–1970	Typhoid fever	4	45	0
	Salmonellosis	4	226	0
	Gastroenteritis	21	5,922	0
1971–1975	Typhoid fever	4	222	0
	Salmonellosis	2	37	0
	Gastroenteritis	63	17,752	0
1976–1980	Salmonellosis	6	1,113	
	Gastroenteritis	114	22,093	0

From Craun, G. F., *Waterborne Diseases in the United States,* CRC Press, Boca Raton, FL, 1986

It is also apparent from these numbers that salmonelloses other than typhoid fever began to appear in the 1940s and in general the numbers of persons affected by waterborne infections of *Salmonella* increased up until 1980. This increase is in spite of the general improvement in water treatment and is counter to our expectations of disease control through water treatment. Perhaps this increase is simply indicative of the overall change in the prevalence of *Salmonella* in the environment, in food source animals, in pets, in domesticated animals, in pest animals, and in the human population. Involved in these

TABLE 3.4
Foodborne Salmonellosis Cases According to Place Where Contaminated Food Was Consumed

Year	Home	Commercial[a]	Other[b]	Total
1950–1960	56(12%)	90(19%)	320(68%)	466
1969	9(19%)	6(34)[c]	22(46%)	47
1970	9(30%)	21(70%)	1(0.03%)	21
1982	12(22%)	33(62%)	8(15%)	33
Totals	86(14%)	160(26%)	351(58%)	596

[a] Includes restaurants, schools, hospitals, and nursing homes.
[b] Includes church, picnics, and social gatherings.
[c] In this number are two cases identified as infected by foods contaminated during processing. These two cases represent only 0.04% of the total reported for this year.

apparent increases, the improvements in reporting within this country must be considered. However, the well-recognized shortcomings, failure to diagnose and to report *Salmonella* infections serve to detract from the appearances of increased numbers.[20]

Failures to diagnose and to report particularly involve salmonellosis (*Salmonella enteritis*, *Salmonella gastroenteritis*) because the vast majority of infections currently being reported are of this type. For example, in the U.S. in 1984, there were 40,861 cases of salmonellosis reported and only 390 cases of typhoid.[15] Most often, salmonellosis begins with a sudden headache, abdominal distress, diarrhea, nausea, and sometimes vomiting approximately 36 to 72 h. after the organisms have been ingested, although the incubation period is commonly said to be 6 to 48 h. This rather lengthy incubation period is an indication that this disease is an actual infection, rather than an intoxication or food poisoning as it is often called. The longer incubation period results because the bacteria must increase in number in the intestinal tract before symptoms of the disease begin. Due to this delay, it is often difficult to determine where contaminated foods were encountered as has been done in the cases reported in Table 3.4. Fever is present in at least 50% of the cases, and when diarrhea occurs, the stools are typically watery and blood specked. Dehydration resulting from vomiting and diarrhea may be severe, particularly in small children.[1]

Some strains have been shown to produce an enterotoxin thought to be responsible for fluid and electrolyte losses from the intestinal tract. This is true of some strains of *S. enteritidis* which produce the cholera-like toxin that may be important in the diarrhea. Increased capillary permeability in the ileum and colon may also account for a part of the fluid loss in diarrhea. Quite often the infection begins and remains an enteritis. It may, however, progress to a systemic febrile infection closely resembling typhoid. This progression is most likely to occur with the paratyphoid serovars, or in some cases, has been reported with enteritidis or typhimurium serovars as the causative agent. In an enteritis or enterocolitis infection, the pathogenic organism is unlikely to be isolated from any specimen other than the feces, but following progression to the febrile illness, isolation may also be made from the blood. These organisms cause a response in which the polymorphonuclear leukocytes (known to kill the bacterial cells rather than to allow multiplication as is the case in the mononuclear response) are increased. In either case, the organism is usually shed in the feces for several weeks (10 to 15% of patients with nontyphoidal gastroenteritis will shed the organism in the stool for 1 to 2 months), but this shedding rarely persists longer than approximately 4 months.[1] No patient, acute or recovering, should be considered to be free of the pathogen until three successive fecal samples obtained three weeks apart result in negative cultures.

There are now over 2000 different serotypes or serovars of *Salmonella* recognized. In the U.S. only about 200 of these are detected or isolated in any given year, and microbiologists in this country are most likely to classify the organisms into three species (*S. typhimurium*, including 1500 serovars; *S. enteritidis*, with an unspecified number of serovars; and *S. choleraesuis*, with only one serovar).[2] The specific serovars which occur most commonly vary from year to year, and from country to country. The World Health Organization[3] (WHO) considers that the *Salmonella* can be classified into three main groups, and that the first of these includes *S. typhi* and *S. paratyphi* A and C. These three serovars are considered to infect only the human and to be spread primarily by either food or water which have been directly or indirectly contaminated by human waste. The second group is considered to include serovars that are host-adapted to other animals. Included in this group are *S. gallinarum* adapted to poultry, *S. dublin* adapted to cattle, *S. abortus equi* adapted to horses, *S. abortus ovis* adapted to sheep, and *S. cholerasuis* and *S. typhisuis* adapted to swine. It is believed that some of these serovars are pathogenic for the human under certain conditions, particularly dublin and cholerasuis. The third group are those organisms which show no particular host adaptation and are pathogenic for either man or other animals. Organisms which cause the majority of salmonellosis infections currently are included in the third group.[3]

It is obvious that the relative importance of the two types of infection have shifted. Because of the high incidence of typhoid and paratyphoid fevers in the past and because of the danger of such infections, there was an intensive study involving the epidemiology, potential complications, transmission, and control of these infections, and an extensive history of the disease has been accumulated. There are some features of these infections, such as the more frequent carrier condition and susceptibility to immunization, that are not seen as often in *Salmonella* gastroenteritis. Most authorities consider that typhoid and/or paratyphoid fevers are caused only by a small number of different organisms or strains of *Salmonella,* whereas the number of different strains of organisms known to cause *Salmonella* gastroenteritis exceeds 2000. It is often thought following diagnosis, that when the infection is produced by a different strain, the disease is different and the same amount or concentration on the history of the infection does not occur. Differences in control of the infections are also apparent in that control of typhoid and paratyphoid in recent years has relied heavily upon immunization and detection and cure of the carrier state. These control measures have not been shown to be effective in helping to contain the transmission of *Salmonella* gastroenteritis, and therefore, other controls must be sought.

Salmonellosis in man, caused by organisms from the second group of organisms, would be considered to be zoonotic salmonellosis cases. In practice, it would also be necessary to consider as zoonoses any infections caused by organisms in the third group which were transmitted from infected animals to the human. Salmonellosis occurs worldwide, but perhaps there are more frequent reports in this country and in Europe than in other parts of the world. This is probably a matter of better diagnosis of infection because the organisms are recognized to cause problems in most areas of the world. Salmonellosis is almost always caused by ingestion of contaminated food, although waterborne transmission, as in the case of typhoid epidemics, is not unknown.

The most generally accepted mode of transmission for salmonellosis has been foodborne transmission; however, there are reports from some areas indicating that this may not be the most common mode, particularly among children. Haddock reports from Guam[21] that among infants and children in that area, contamination of soil in which the children play and spread by aerosol from patients may be involved as major modes of transmission. Regardless of the mode of transmission, ingestion of the organisms provides the infectious dose of cells which grow in the intestinal tract. With this growth, inflammatory lesions develop in the lining of the intestinal tract, and the organism rarely invades deeply enough to produce a bacteremia or septicemia. In more recent outbreaks in this country which were found to be caused by *S. enteritidis* (SE), the symptoms include diarrhea, vomiting, abdominal pain, chills, fever, and headache. It has generally been found that many of the SE infections in recent years have had considerable seriousness as evidenced by the numbers of deaths which occurred.[22] In studies in England, over a 20-year period of over 7000 cases of *Salmonella* infections, enteritis occurred in 68%, and septicemic infection or enteric fever occurred in the remainder.[1] The highest fatality rates were observed in the enteric fever manifestations and in the elderly and very young. Death rarely results from the enterocolitis type of infection, except in cases of dehydration in the very young, in the elderly, and in those individuals who have some other underlying disease. Enteric fever infections still result in the greatest fatality rates.

While one might expect that with improved sanitary and living conditions in the western world, the incidence of salmonellosis would have decreased, the opposite trend has been observed in the U.S. since 1945. Whether this apparent increase in infection rate is real or is simply the result of better diagnosis and reporting is open to question. Certainly the public is more aware of the likelihood of acquiring a gastrointestinal infection from contaminated foods, and it is entirely possible that this awareness stimulates increased consultation of physicians, improved diagnosis, and better reporting so that we are simply getting closer to the actual numbers of infections, rather than being misled with unrealistic numbers as we may have been in the past.

Another possible factor in the apparent increase in incidence of salmonellosis is the change in food habits, particularly for the human, but also for pets, and especially in the U.S., although a similar change has occurred in the Western world generally. The public now consumes more processed foods, and consumes more food away from home that has been prepared by others. The general public now also uses more processed animal foods for pets and other domestic animals. It is therefore obvious that there have been more chances that the foods consumed by the human or pets could have been contaminated before consumption. Since the 1960s when the largest increases in incidence of salmonellosis began to appear, it has become more and more apparent that foodborne salmonellosis presents a public health problem. In the 1960s, discussions began among health and food authorities which were intended to control the transmission of *Salmonella* to the human.[4]

Recently, it has been grossly understated that "nearly all food scientists agree that the salmonella bacteria . . . is proving difficult to control."[23] A number of factors are recognized as contributing to the difficulties in control of these organisms. Some of the factors involve the animal hosts of the bacteria, some the nature of the bacteria themselves, and some the processes used to produce, process, distribute, and serve foods to the public.

Because of publicity received, perhaps too many people consider that the single, major source of *Salmonella* causing infections in the human is contaminated poultry. It is true that poultry, primarily chickens and turkeys, are frequently carriers or are infected with *Salmonella* of some types. Depending upon who is reporting, the percentage of poultry contaminated with *Salmonella* varies up to 64% when the poultry reaches the consumer,[23] although it is impossible to find accurate figures on the actual rate of infection or infestation in poultry flocks, or of contamination of processed poultry carcasses on the market with *Salmonella* over a long period of time.

Some authorities state that most poultry products are contaminated with *Salmonella* species.[24] On the other hand, recent data from our laboratory indicate that the amount of poultry on the market contaminated with these bacteria may actually be reduced rather drastically in recent years.[25] Because of the methods used in processing poultry, the actual incidence on the processed meat of pathogenic bacteria in general is no doubt much higher than the incidence within the flocks prior to shipping and processing. Each carcass is likely to have low numbers of *Salmonella* on the surface, if contaminated in this manner, but these are sufficient to contaminate the handlers and kitchens of the consumers, where the organisms may grow and provide sufficient numbers for the infectious dose for the consumer. Since poultry carry most serovars with slight or no symptoms, the infection can remain undetected in flocks, and the animals may be shipped to the processing plant where the procedures used may almost ensure that the contamination of one bird will be transmitted to others.

This statement is not intended as criticism of procedures used in poultry processing plants, although these have come under heavy criticism in recent months.[24] Those procedures are designed, inspected, and controlled by governmental authorities and specialists in the industry and are based on recommendations which have a basis in studies intended to detect the methods most likely to create the least problem for the processor and the public. The poultry industry and others have worked hard to improve the situation of contamination of processed poultry products through the NPIP.[26] In order to maintain reasonable costs for the consumer, with the best possible conditions for the food product, the best procedures must be used by the processors. Processing methods for poultry are frequently reviewed by international groups, industry representatives, and governmental authorities.[3] From these reviews, come improvements in methods which help to reduce the spread of contamination, and therefore, reduce the spread or transmission of *Salmonella* for the outbreak of infections. When processing is carried out correctly, the number of bacteria contaminating the surface of carcasses is small. Therefore, it is necessary that the conditions be rigidly controlled so that the numbers of bacteria on surfaces of marketbound poultry remain small during handling, shipping, and service preparation. A major factor during this time is the maintenance of proper temperatures in any environment where the poultry is located. Parmley[27] reported that during processing, bacteria which contaminate chicken skin easily enter crevices in the skin surface as well as feather follicles. Washing at that stage will not remove these bacteria, whether *Salmonella* or other types, and each time the carcass is dipped into the water the bacteria may enter a little deeper into the crevices. Two methods are under consideration to help to reduce these problems. Spray rather than immersion, and if immersion is to be used for chill bath, then the carcass can be wrapped in plastic bags prior to chilling, in order to keep the chickens from trading bacterial flora and to prevent the deeper entrance of contaminating bacteria into crevices and follicles.

Altekruse has reported that several variables affect the prevalence of disease in flocks, but that biosecurity measures influence the extent of the disease in flocks.[28] He further states that ovarian infection is present in roughly half of affected hens, and that the hen is affected in egg production in severe infections. The poultry may also exchange contaminating pathogenic organisms while in the cages or boxes in which they are placed for shipment to market. It is believed by some authorities that shipment itself creates stress in the poultry, resulting in increased shedding of bacteria present in the gut. This shedding then provides additional opportunity for exchange of contamination during the shipment process. In a recent report, Moran and Bilgili[29] state that the susceptibility of uninfected broilers to *Salmonella* has been shown to increase with the stresses of feed and water removal, crating, and hauling. These stresses increase voiding of the cecal contents during haul, and therefore, cross-contamination may ensue directly because of contact between birds.

In addition to enteric fever and enterocolitis types of salmonellosis infections, focal infections have also been reported in most organs of the human body. Many focal infections are considered likely to have resulted from a bacteremia involving the organisms, and it is thought that some occur because there have been prolonged or untreated cases of enterocolitis which have allowed the organisms to be transferred directly or indirectly from the gastrointestinal tract to the organ involved. Focal infections have also been observed to occur more frequently when there is some underlying disease, particularly disease involving the organ in which the salmonellosis localizes. In chronic focal infections, only local symptoms may appear, and fever may not be present. In acute focal infections, treatment is essential and should be accomplished with rapidity. Often these infections are difficult to treat because of multiple antibiotic resistance in the organisms causing the disease. Prognosis is poor when the focal infections occur in the heart or the central nervous system regardless of the speed and type of treatment.

The methods for culture of all *Salmonellas* were originally developed to detect these pathogens in clinical specimens. When an organism is causing an infection in the human, it is likely to be present in relatively large numbers at the site or location of the infection. This simply means that it is not as likely to be overwhelmed or overgrown by other organisms present in the same environment, and therefore enrichment procedures may not be as essential as when these numbers make up only a small portion of the microbial populations as is likely to be the case in contaminated foods. In the event of outbreaks of infection, it often will become necessary that clinical specimens, food specimens, and environmental specimens be collected and cultured. In these cases, the collection of the proper specimen in the right way, and subsequently, the proper shipment of that specimen to the laboratory, may well be as important as the correct culture media and procedures.[30] Laboratory culture procedures and culture media may vary somewhat for clinical specimens and should be determined by the clinical or public health laboratory handling the work for the investigation. Laboratory methods, and media most useful for foods and environmental samples, have been tested and approved by the Food and Drug Administration (FDA) and are discussed in Chapter 8.

To diagnose salmonellosis, blood cultures are often useless because they remain negative, but fecal cultures are usually positive within the 1st week and may remain so for several weeks — even after clinical recovery is complete. If the infection is suspected to be a case of salmonellosis, it is well to use enrichment cultures because early fecal cultures may not show positives. It is possible that the *Salmonella* present will be in small numbers and may be consistently overgrown by other enteric bacteria present in the bowel. Fecal specimens should be immediately plated to selective and differential media, and cultures should be examined for the typical colonial growth expected for *Salmonella* strains. For uncomplicated enterocolitis in an otherwise healthy adult, no specific treatment other than rehydration and electrolyte replacement is usually indicated, contrary to the needs in treatment of typhoid. Antibiotics used in salmonellosis infections may actually prolong the carrier state and also may result in production of resistant strains of bacteria as has been observed when antibiotics are included in animal feeds. In infants, however, as well as in the elderly, and in others who have a coexisting disease, antibiotic therapy should be used. Ampicillin or amoxicillin are usually effective, and if not, then chloramphenicol or trimethoprim-sulfamethoxizole should be used to treat. The latter two drugs may also be useful when strains show resistance to other antibiotics.

REFERENCES

1. **Roberts, R. B.,** *Infectious Diseases. Pathogenesis, Diagnosis, and Therapy*, Year Book Medical Publishers, Chicago, 1986.
2. **Burrows, W.,** *Burrows Textbook of Microbiology*, 22nd ed., Revised by Bob A. Freeman, W. B. Saunders, Philadelphia, 1985.
3. **WHO Expert Committee,** *Salmonellosis Control: The Role of Animal and Product Hygiene*, Tech. Rep. Ser. No. 774, World Health Organization, Geneva, 1988.
4. **Safe Drinking Water Committee,** *Drinking Water and Health*, National Research Council, National Academy of Sciences, Washington, D.C., 1977.
5. **Hornick, R. B., Greisman, S. E., Woodward, T. E., DuPont, H. L., Dawkins, A. T., and Snyder, M. J.,** Typhoid fever: pathogenesis and immunologic control, *N. Engl. J. Med.,* 283, 686, 1970.
6. **Hockin, J. C., D'Aoust, J. Y., Bowering, D., Jessop, J. H., Khanna, B., Lior, H., and Milling, M. E.,** An international outbreak of *Salmonella nima* from imported chocolate, *J. Food Prot.,* 52(1), 51, 1989.
7. **Craven, P.C., Mackel, D. C., Baine, W. B., Barker, W. H., Gangarosa, E. J., Goldfield, M., Rosenfield, H., Altman, R., Lachspelle, G., Davies, J. W., and Swanson, R. C.,** International outbreak of *Salmonella eastbourne* infection traced to contaminated chocolate, *Lancet,* 1, 788, 1975.
8. **D'Aoust, J. Y.,** Infective dose of *Salmonella typhimurium* in cheddar cheese, *Am. J. Epidemiol.,* 122, 717, 1985.
9. **McCullough, N. B. and Eisele, C. W.,** Experimental human salmonellosis. I. Pathogenicity of strains of *Salmonella meleagridis* and *Salmonella anatum* obtained from spray-dried whole egg, *J. Infect. Dis.,* 88, 278, 1951.
10. **McCullough, N. B. and Eisele, C. W.,** Experimental human salmonellosis. III. Pathogenicity of strains of *Salmonella newport, Salmonella derby* and *Salmonella bareilly* obtained from spray-dried whole egg, *J. Infect. Dis.,* 89, 209, 1951.
11. **Matic, S., Mihokovica, V., Katusin-Razem, B., and Razem, D.,** The eradication of *Salmonella* in egg powder by gamma irradiation, *J. Food Prot.,* 53, 111, 1990.
12. **Blaser, M. J. and Newman, L. S.,** A review of human salmonellosis. I. Infective dose, *Rev. Inf. Dis.,* 4, 1096, 1982.

13. **Xu, H. S., Roberts, N. C., Adams, N. B., West, P. A., Siebeling, R. J., Huq, A., Huq, M. I., Rahman, R., and Colwell R. R.,** An indirect fluorescent antibody staining procedure for detection of *Vibrio cholerae* serovar 01 cells in aquatic environmental samples, *J. Microbiol. Methods,* 2, 221, 1984.

14. **Cruickshank, R., Ed.,** *Medical Microbiology,* Williams & Wilkins, Baltimore, 1965.

15. Annual Summary, 1984, Morbidity and Mortality Weekly Report, **Centers for Disease Control,** Public Health Service, U.S. Department of Health and Human Services, Atlanta, GA, 1986.

16. **Littman, A., Vaichulis, J. A., Ivy, A. C., Kaplan, R., and Baer, W. H.,** The chronic typhoid carrier. I. The natural course of the carrier state, *Am. J. Public Health,* 38, 1675, 1948.

17. **RosenKrantz, B. G., Advisory Ed.,** *The Carrier State,* Arno Press, New York, 1977.

18. **Sanders, E., Brachman, P. S., Friedman, E. A., Godsby, J., and McCall, E.,** Salmonellosis in the United States. Results of nationwide surveillance, *Am. J. Epidemiol.,* 81(3), 370, 1965.

19. **Fox, J. L.,** In search of a better Typhoid vaccine, *ASM News,* 54(10), 551, 1988.

20. **Craun, G. F.,** *Waterborne Diseases in the United States*, CRC Press, Boca Raton, FL, 1971.

21. **Haddock, R. L.,** Three modes of Salmonella transmission, in *Proceedings of the 2nd International Symposium on Public Health in Asia and the Pacific Basin,* Muangman, D., Ed., Vol. 2, Mahidol University, Bangkok, 1986.

22. **Blumenthal, D.,** *Salmonella enteritidis.* From the chicken to the egg, *FDA Consumer,* U.S. Department of Health and Human Services, Washington, D.C., April 1990, 7.

23. **Guthrie, R. K.,** Unpublished data from report to Channel 11 television station, Houston, TX, 1987.

24. **American Society for Microbiology,** *ASM News,* 55(10), 531, 1989.

25. **Guthrie, R. K.,** Unpublished laboratory data, 1989.

26. *Federal Register,* 9CFR Parts 145 and 147, Vol. 54, No. 106, Rules and Regulations, Animal and Plant Health Inspection Service, U.S. Department of Agriculture, 1989.

27. **Parmley, M. A.,** Researching the Salmonella problem, *Food News for Consumers,* 5(2), 8, 1988.

28. **Altekruse, S.F.,** Clinical Presentation of *Salmonella enteritidis* Phage-Type 4, Report at *Salmonella enteritidis* Regional Work Conference Central Region, Kansas City, MO, 1990.

29. **Moran, E. T., Jr. and Bilgili, S. F.,** Influence of feeding and fasting broilers prior to marketing on cecal access of orally administered *Salmonella, J. Food Prot.,* 53, 205, 1990.

30. **International Association of Milk, Food and Environmental Sanitarians, Inc.,** *Procedures to Investigate Foodborne Illness,* 4th ed., International Association of Milk, Food and Environmental Sanitarians, Ames, IA,1987.

Chapter 4

CONTROL OF *SALMONELLA* SPREAD

The *Salmonella* are natural inhabitants of the intestinal tract of food animals, including cattle, swine, sheep, chickens, turkeys, and ducks, and of pet animals including dogs, cats, and turtles. *Salmonella* are also found in wild animals of all types. Pigeons, reptiles, mice, rats, and insects present special problems involving environmental sanitation, because they are all susceptible to infection and live essentially uncontrolled. As intestinal inhabitants in so many animal species, salmonella are shed by infected animals in fecal material, and in many cases, in urine as well. From these sources, the bacteria contaminate the environment generally and can grow in foods, waters, and on inanimate objects which are contaminated by feces for periods of hours or days and often come in contact with other susceptible animals in whom a new infection is begun. Salmonella are mesophilic bacteria, and as such grow best at a temperature between 15 and 40°C although the range is said to be 8 to 45°C.[1,2] This characteristic also means that the bacteria grow well at warm room temperatures, and thus are able to increase in numbers in foods when contamination has occurred. Many persons lose sight of the fact that with the short generation time of salmonella, a matter of minutes when in a good culture medium such as most foods, a contaminating population in the kitchen may increase significantly within an hour's time. Under less ideal conditions, the salmonella will still double populations, albeit in a longer time. Most salmonella begin to die off at a temperature slightly above 60 °C; however, some foods, particularly those high in fat and low in moisture such as chocolate, appear to offer the bacteria some protection against heat, and in such a medium, the organisms may not begin to die off until the temperature reaches more than 70°C, and even then a considerable time may be required to kill off large numbers of bacterial cells. Factors involved in susceptibility or resistance to heat of bacteria include, in addition to the degree of heat, the type of organism, the numbers of cells, the stages of growth of the cells, the temperature at which the organisms were grown, and the medium on which they were grown as well as the medium in which the bacteria are heated. At low or refrigerator temperatures, the organisms do not grow rapidly; in fact the generation time may be 8 to 12 h. Again, depending on the same general factors as above, even at these temperatures, the organisms will continue to grow slowly, with a lag phase of several hours. In foods, growth may be either slower or faster than the times above which were measured in various culture media. It appears that some food types provide protection to the organisms against adverse conditions in the same way that some foods appear to allow salmonella to produce infection in the human from smaller initial numbers. When conditions are somewhat less favorable, as in dry or very cold environments, the organisms remain viable, but do not always multiply.[3] The salmonella, like many other pathogens, can be virtually eliminated from various environments, including foods, by the use of ionizing radiation. Radiation used in this way is measured in rads or krads (dose of radiation absorbed) and there is much discussion about the use of radiation as a means of food preservation at the present time. The Food and Drug Administration (FDA) has not approved the general use of radiation for food preservation; however, there is some use of this method, particularly in the armed forces. Doses on the order of 250 to 500 krad of radiation are needed to wipe out populations of salmonella, and spore-forming organisms will require much larger doses.

Bacteria, in general, require the presence of water in usable form for growth. The measure of water availability which is most often used in reference to foods is water activity ($_{Aw}$). $_{Aw}$ in a food may be increased either by adding solutes or by removing water. The actual definition of water activity of a food is the ratio of the water vapor pressure of the food to that of pure water at the same temperature. Thus, when a solution becomes more concentrated the vapor pressure decreases, and the $_{Aw}$ falls from a maximum value of 1.0 for pure water to a fraction of this value. The salmonella need an approximate 0.95_{Aw} for growth at near optimal temperatures. Values below this will restrict or stop growth of these organisms.

The fact that these bacteria are normally found in the animal body, has led in the past to the supposition that, at least some strains are host-restricted, and this has been used in naming strains or serovars. It appears that the serovars typhi and paratyphi A are restricted to the human, as is paratyphi B in most instances, but other serovars seem to be much less restricted as to the animal species that can be infected.[4] A wide variety of serovars are commonly found in this country, as in others, with the most common strains varying year to year. Most likely, with increases in travel, the exchange of strains between geographical localities will increase. For some of the *Salmonella*, the serovars have been further subdivided either on the basis of antigenic differences, or on the basis of phage susceptibility. Most recently, this subdivision has been the basis for restrictions on the movement of poultry and poultry products into this country from Europe of a particular phage type of *S. enteritidis* (Phage type 4).[5]

Although the salmonella naturally inhabit the intestinal tract of animals, this does not mean that they do not survive, and therefore do not continue to live, in other environments. These bacteria are not found in natural habitats outside the animal host, but are able to survive and as indicated above, even to grow. In some cases, the organisms appear to survive for long periods in the environment outside of the animal body; however, it is sometimes difficult to determine whether the organisms have survived for long periods in the environment, or whether the environment has simply been recontaminated from infected humans or animals.[6] One of the characteristics of these bacteria, like other enterobacteria, is that they are very resistant, and very adaptable to environmental conditions. The organisms are able to grow over a wide range of temperatures, and in a wide variety of nutrient dilutions and pH conditions; however, many of the gram-negative bacilli, including the *Salmonella,* are more sensitive to freezing than are the gram-positive bacteria. In some foods, however, some of these organisms appear to be protected from the damage of freezing. For example, *S. typhimurium* in chow mein survived with a 20% recovery after storage at −25°C for a period of 9 months. Other salmonella species have been shown to have survived in ice cream for a matter of years.[7] Since the salmonella are frequently reported to have been isolated from frozen foods of different varieties, it can safely be assumed that just the use of freezing temperatures is not any protection from transmission of salmonella to the susceptible individual. From feces and urine, the organisms contaminate soils, from which they are washed into surface waters, and are able to survive, and in some cases grow for considerable periods of time. Resistance of the bacteria to drying permits them to survive in dust as well as dried fecal and urine stains for a matter of weeks to even months. They thus remain available for contamination of water and soils from which they are again able to gain entry by some means into the intestinal tract of animals. Generally, it is accepted that the enterobacteria are all allochthonous organisms when present in either soil or water. Whereas autochthonous or indigenous organisms occupy a niche in an environment on a permanent basis, such pathogens as salmonella, which are adapted to the intestinal tract, are not able to be established in all environments as consistent residents. In the intestinal tract, as pathogens, they are frequently the producers of symptoms of disease, and the resistance mechanisms of the animal are aimed at their removal. In the environment, since they do not occupy a permanent niche, they are unable to survive indefinitely, although the more resistant the organisms are to environmental factors to which they are subjected, the longer they are able to survive in soil and water. The longer the bacteria survive in these environments, the better the chance that they will eventually be transmitted to an animal in which they are able to set up an infection. It is from these sources that other animals are most likely to be infected and foods which are being processed and handled will be contaminated, and the organisms will thereby be able to transmit the infection to the susceptible human.[8-10]

The intestinal shed of salmonella by all animals infected, including those not showing symptoms of infection, presents a number of problems in environmental sanitation which are critical to the production and processing of foods not usually contaminated by these organisms. For example, once an animal production facility has been occupied by herds or flocks for any period of time, it must be assumed that such housing is contaminated with the pathogenic microorganisms even though it is accepted that the bacteria can not survive indefinitely. Once farm animals are infected with the bacteria, they are not treated in such a way to completely kill the infection in all animals on the farm, and during the infection, the bacteria are constantly shed in fecal matter. Those bacteria which are shed, contaminate the production facilities, and remain to be spread from the environment to additional susceptible animals. This is true of sties housing swine; farms, pens, and barns housing cattle; poultry houses; and the open areas where these animals are allowed to range. It is essential, if one hopes to establish any type of salmonella-free animal herd or flock, that a number of stringent measures be taken to assure that the bacteria are not present in the animals or the facilities, or if they are present, that they are removed. The production facility must be established as a salmonella-free environment, and nothing but salmonella-free breeding stock must be housed in these facilities.[10,11] For this effort to be successful, all animal shelters or facilities must be constructed of smooth, long lasting, easily washed and disinfected materials which can be subjected to disinfection and fumigation to kill any organisms which contaminate. Such facilities must be constructed so that fecal material and urine from animals will be collected and removed constantly and so that these waste materials can be treated to kill any pathogenic organisms which are present. Openings into the facility must be controllable so that entry of persons, animals, or objects can be controlled, and so that aerosolized contamination can not enter. Animals brought into such a facility must be certified salmonella-free and must be brought in on an all-in, all-out basis so that herds or flocks are not mixed once the salmonella-free status is achieved. This type of operation is a part of the regulations used by the SE Task Force discussed in Chapter 7. The entry of all materials into the facility must be controlled so that contamination of the animals will not occur from feed, a very common source of salmonella contamination for poultry flocks, nor from equipment or water. Experimental maintenance of food animals has been successfully accomplished in Denmark, Sweden, and the Netherlands, and therefore, the world has been shown that such control is possible.[11]

The cost of maintaining totally pathogen-free facilities in the production of food animals has not yet been assessed on anything like a national or world scale. The cost of this production method will certainly be passed on to the consumer, and therefore it must be at an acceptable level if the consumer continues to use the food product. Production of *Salmonella*-free meats is useless unless these are accepted by the consuming public, and these will be accepted only if the cost is deemed reasonable.

To begin to appreciate the costs which would be associated with establishing a *Salmonella*-free environment consider what must be done to assure such a condition for a commercial poultry production unit. And to appreciate the importance of such maintenance the fact must be accepted that without the maintenance of a *Salmonella*-free environment the spread of the organisms to susceptible food animals, and on to the human can not be controlled. In their current disease prevention program for *Salmonella enteritidis*, the U.S. Department of Agriculture (USDA) is attempting to direct the establishment, and the continued maintenance of *Salmonella*-free environments. The premises must be controlled by construction of pens, houses, etc., which can be maintained in such a way as to exclude contact of the poultry with wild birds, rodents, household pets, and wild animals.[12-14] If the environment is controlled as well as possible, then it becomes even more important to control the presence of the infection, or the pathogenic organisms in the animal flocks. A number of investigators have begun to work to develop methods of preventing, or controlling the numbers of pathogens in domestic animals, and at the least to control the transfer of pathogens to imported, or newborn animals. Schor[15] reports that addition of lactose to the drinking water of newborn chicks blocks infection with *S. typhimurium*, one of the common infective strains, at a cost of approximately 0.5¢ per bird. Other studies[16] have shown that giving cecal contents from adult, salmonella-free chicks to young chicks at under 1 day of age will protect the chicks from later salmonella infection. These methods need to be more fully studied and confirmed, and the practice of protecting chickens from salmonella infection can be promoted for helping to control the spread of these pathogens by way of contaminated poultry.

A major problem in producing poultry which is salmonella-free is assurance of avoidance of contact of wild birds with the poultry flock. This problem must include the prohibition of any bird, wild or pet, from reaching the environment, including the feed and water, of the poultry flock. The building housing the operation must be bird-proofed, as well as rodent-proofed. The reason that no contact between these animals and the poultry can be permitted, is that birds of all types are frequently contaminated, or infected with some serovar of salmonella. The construction of the facility must be such that it can be of a nonwettable, smooth material so that cleaning and disinfection can be carried out efficiently. Floors and lower walls should be of concrete, and should be continuous to avoid seams, and floors should slope to a central flushing drain to adequately remove wastes and allow for proper cleaning and disinfection. The building should be constructed to be rodent-proof, and should provide large areas outside of doors as concrete slabs without seams. This slab will help to reduce the danger of tracking in salmonella on the feet of workers. Even more efficient is the maintenance of a boot-bath of disinfectant of adequate strength to overcome any contamination on the feet of the workers.

Since the production facility must be protected against dustborne contaminants, this necessitates controlled air access. Such facilities are generally not considered for commercial production operations; however, at least control of direct contact of the animals with outside dirt, dust, and fecal materials from other animals should be practiced in limiting access to the production facility.

If production facilities are maintained as salmonella-free environments, and then feeds brought into the facilities are not checked for the same condition, then both time and money have been wasted. It is often said that much animal contamination comes from feeds used in such production facilities. To avoid food transmission of salmonella infections to food animals, it is essential that animal feeds be produced under the same stringent sanitary standards required by Good Manufacturing Practices (GMP) for human food.[17]

FIGURE 4.1   The housefly and rodent are the more common pests which have been shown to be involved in transmission of *Salmonella typhi* and other types of *Salmonella* to the human.

Generally, good standards are required for the production of animal feeds, although the stringency is usually nothing like requirements for human food. One of the cost reducing steps taken in animal feed production to avoid spread of pathogenic organisms, as well as to increase the growth rate and general state of health of the food animals, is the addition of antibiotics to the feeds. This practice has been criticized as one of the factors in the production of antibiotic resistant strains of *Salmonella*; a problem which will be discussed in more detail in Chapter 5.[18] For production of feeds which are free of pathogens, the single most important factor must be the use of raw materials of good bacteriological quality. To assure this end, the raw materials must be bacteriologically tested to rule out the presence of salmonellas, or other pathogens, and then must be maintained in this condition. Many vegetable meals, and seeds which are used in production of poultry feeds are initially contaminated with salmonella or other pathogens. If this is the case, the raw materials must be decontaminated before use in the finished feed. Raw materials, all ingredients, and all stages of production of the finished feed must be protected from contamination by pests, insects, birds, or rodents (Figure 4.1), and stored in a dry place free of dust (the same is true of any human foods being processed). In the manufacture of feeds for some animals, the feeds can be combined, blended, and processed to produce pellets or cubes which can be treated by heat during processing. This practice is not commonly used for poultry feeds since these animals are unable to chew the pellets or cubes during ingestion. Since irradiation is now allowed for the preservation of some human foods, this is a method of decontamination for animal feeds which shows good promise for the future. It has the advantage that it can be carried out after the feed is packaged to avoid recontamination, and it does not change the texture or the food value of the product. A disadvantage of irradiation as a means of decontamination is that a relatively long exposure time is required (up to 24 h) for inactivation of the pathogenic organisms. Once the feed is produced in an uncontaminated state, care must be taken to assure adequate storage to avoid recontamination from either physical environmental factors or from contact by birds, rodents, insects, or other animals. To assure the uncontaminated state, the finished products must be tested, in at least spot checks, and bacteriological cultures must be used to assess quality control for the products. In this case, as in environmental testing, it would be necessary to use the fecal pollution indicator organisms to assess the probability that all pathogens have been removed from the products.

The assurance of salmonella-free water supplied to poultry flocks is more difficult than is the case for feeds, only from the standpoint of maintenance of the water in a salmonella-free state. If water is obtained from an approved water supply, as is available in most locations in this country, then it should prove to be salmonella- and even pathogen-free.[8] When it is supplied continuously to the poultry flock, it is much more difficult to maintain in this state, because it must be left open and available to the animals on a 24-h basis. If it is kept in this manner, then it is open to air, dust, the poultry themselves, and all manner of contamination. Unlike dried feeds, if the water is contaminated, and is left for any period of time, it is possible for some bacterial growth to take place because there will likely be some organic contaminants in the water which can be utilized as nutrients for the bacteria during growth.

The introduction of salmonella into surface waters is very difficult to control. Since the organisms are shed in the fecal matter from every animal infected, whether symptoms of illness are present or not, these bacteria will reach water supplies on the earth's surface either from sewage effluents, or in surface waters from other, nonpoint polluting sources. It is generally agreed that salmonella do not multiply in relatively clean water, but, if the water is polluted, and there are adequate nutrient resources present, in the presence of suitable pH and temperature there will be some slow growth or multiplication of these organisms. This is true of all those organisms other than *S. typhi* which is adapted to the human host, and which is not generally believed to reproduce outside the human body, although it will remain viable for considerable periods of time. Even if the bacteria suffer injury, and are more difficult to culture, it is most likely that the salmonella will be culturable from almost all sources, because the culture procedures for these bacteria generally include pre-enrichment and enrichment steps, which allow recovery and growth of the bacteria under most circumstances.[19] Although waterborne outbreaks of salmonellosis are sometimes reported, it is unlikely that consumption of water from approved sources in the U.S. will result in the production of infection of this type, and this mode of transmission appears to be relatively unimportant.[20]

Salmonella are often introduced into an environment by persons who work or visit there.[21] Because of this, it is essential that this contamination be taken into account by persons attempting to maintain salmonella-free environments in the production of food animals. Although our major consideration is to attempt to prevent passage of contamination from other animals or the environment to the human, it must be admitted that once contaminated, the human is then often a continuing source of organisms which are spread to the environment. Because of this, everyone who comes in contact with the production environment in any capacity, i.e., owner, worker, delivery person, salesperson, buyer — anyone — must be assumed to be a potential source of contamination, and must be treated as such. Traffic patterns must be controlled, and contact with animals, feed, water, or equipment must be carefully monitored to prevent introduction of a salmonella serovar into the production facility. Visitors should be severely limited, if not prohibited, and transient workers should be kept to a minimum to control contamination. Direct passage from building to building by employees must be avoided, or at least precautions for disinfection, especially of shoe soles, must be taken between buildings when such passage is required. The avoidance of direct passage is as critical to control of contamination as is avoidance of direct passage from patient to patient by medical personnel in a hospital setting. Employees and visitors alike should wear protective outer clothing and rubber disinfected overshoes in movement around a production facility. These coverings are more for the protection of the environment and the birds than for the clothing of the persons.[8]

In the maintenance of salmonella-free production facilities, birds, or any animals of the same age should be acquired from one source and moved into a facility all at one time. This is the initial step in the all-in, all-out principle. The temptation to mix birds from different sources or of different ages should be resisted to allow ample time for testing to show that all the animals are clean, and that there is no contamination coming from one or the other of the sources of supply. Mixing of species of birds or animals should also be avoided to reduce the chance of contamination spread on any farm. In this vein, production facility buildings should be dispersed on a farm so that spread of contamination from one to another can be reduced, and adequate obstacles can be maintained to obstruct spread.

In animal production facilities, it is critically important that sick or dead animals be removed from the facility and that disposal be carried out efficiently as rapidly is possible. Incineration is usually adequate for disposing of carcasses; however, sealed decontamination pits may also be used satisfactorily. If animals are to be collected by a commercial operation for removal, it must be remembered that the personnel involved in the removal will likely be very good possibilities as sources of salmonella or other pathogen contamination. It is best to assure that such persons have the minimum exposure to the production facility, and that decontamination of the environment be accomplished as quickly and efficiently as is possible. This means that the buildings must be cleaned (minimizing the raising of dust), disinfected, and that all litter must be removed, stored to avoid spread of contamination, and disinfected.

One feature of salmonella contamination in poultry flocks, and in other food animals as well, is that not all animals will show symptoms of the presence of the salmonella. Frequently, animals carrying and shedding salmonella organisms, appear perfectly normal, and spread of contamination can be detected only when it appears with symptoms in another animal, or when it is detected by microbiological procedures. Testing programs for poultry production flocks are not carried out on a routine basis, but do occur in certain emergency situations when it is important in the control of the transmission of salmonella serovars to flocks that are susceptible, or that are in production for breeding or egg-laying purposes. In the case of sick or dead animals, it is critically important that microbiological procedures be instituted as soon as possible for the detection of contamination. In any event, when sick or diseased animals have been detected in a production facility, it is best to completely remove all the animals, and to disinfect the facility before more animals are brought in. It is sometimes recommended that the production facility be left empty for some weeks if possible before other animals are brought in.[22,23] This allows additional time for the disinfection and decontamination measures to have maximum effect, and frequently may help to avoid contamination of the new animals brought in.

A major consideration in cleaning up a production facility after sickness or disease has been detected in food animals is the dismantling, cleaning, and disinfection of any equipment present in the facility. It is of little use to wash down and disinfect walls and floors if cages, racks, and feeding utensils are not cleaned and disinfected as thoroughly. This cleaning process should be done with high-pressure hoses after equipment is dismantled, and must be followed by disinfection with chemicals of adequate strength to accomplish killing of any pathogens. One of the most critical factors in such cleaning is that adequate time be allowed for chemical disinfectants to act after application. In all chemical disinfection, the three critical considerations are the nature of the chemical, the amount of dilution used, and the time allowed for the chemical to act.[8]

Different types of facilities and equipment used for the production of food animals will require different chemicals and processes for cleaning, disinfection, and for assuring removal of potential pathogens from the environment.[24] Animal houses, whether for poultry, swine, cattle, or other types of food animals must be thoroughly cleaned whenever emptied to provide the opportunity. This cleaning must include the removal of all litter, manure, left over feed, and in so far as possible, the dust resulting from any and all of these. Then the surfaces must be hosed down, and left wet for at least 24 h in order that any soiled materials which have dried on surfaces may be removed. Following this wet period, the remaining materials can be best removed by the use of high-pressure hoses which provide a generous supply of cold water. After thorough cleaning is completed and assured, then the disinfectant chemicals can be applied. A most critical factor at this time is that all disinfectant chemicals be allowed to remain, and to act for a sufficient time. For most purposes, 1-h applications of disinfectant chemicals are sufficient. In disinfection of animal houses, the successive use of a 3% solution of sodium hydroxide at a temperature between 70 and 80°C; a 2% solution of formaldehyde (use of this chemical is now restricted by recent federal regulations) at a temperature between 25 and 30°C; and a 2% chlorine content in calcium hypochlorite at a temperature between 15 and 20°C are effective. Each solution should be allowed to act for 1 h before the next is applied. Each chemical disinfects, and, therefore, these time intervals simply allow time for the action of each to be effective. The use of formaldehyde is now restricted in the U.S. by regulations promulgated by the Occupational Safety and Health Administration of the federal government. In many European countries, however, it is recommended that this chemical be used as a disinfectant, and in fumigation.

Final disinfection of such premises is most effectively accomplished by the application of fumigation using formaldehyde (30 ml liquid formalin at 40% strength, and 20 g of potassium permanganate for each cubic meter capacity of the facility). Such fumigation will be effective only if all surfaces are wet, and the temperature is around 15°C.

Equipment used in the animal houses should be scrubbed with the same chemicals used to wash down the facility, and should be brought back inside before fumigation is carried out so that they will also be subjected to the disinfectant action of the fumigation. It is impossible for most production facilities to sterilize materials used for litter (straw or wood shavings), so it therefore is essential that these materials are obtained only from sources which can assure the user that they have not been subjected to contamination through contact with animals of any kind.

Animals and meats can not be subjected to fumigation or disinfection to prevent the presence of contamination; however, one product which is often contaminated can be fumigated to reduce the risk of contamination. This product is the poultry egg (Figure 4.2). Any bacteria present in the intestinal tract of the poultry will be transmitted to the shell of the egg in the passage of the egg through the canal. As the egg cools, bacteria which are present on the surface may be allowed to penetrate the porous shell and membrane, and once this has happened, fumigation will not be effective. The best practice then is to collect the eggs as soon as possible after they are layed, and to subject them to fumigation while they are still warm. The same chemicals are used in the fumigation of eggs, as are recommended for fumigation of facilities, although some special precautions must be taken to assure that the fumigant reaches the desired surfaces. The eggs should be placed in wire baskets with ample openings so that fumigant can pass through, and baskets should be placed in cabinets in such a way as to allow ample space for air and fumigant circulation. The cabinet temperature must be maintained between 20 and 25°C and air circulation must be assured to circulate the fumigant throughout the cabinet. The cabinet must be maintained in high humidity to assure the

FIGURE 4.2   Although the porous shell allows entrance of bacteria into the egg, the same characteristic permits efficient fumigation to kill contaminants.

best action of the fumigant. Again a 1-h exposure should assure adequate disinfection. Once fumigated, whether the eggs are to be used for hatching or for shipment to market for consumption, care should be taken to assure that the eggs are placed into containers which are not contaminated, and preferably into containers which have also been fumigated.

Chemical disinfectants which are likely to be absorbed by foods, and thereby to be termed adulterants, can not be used for premises used for animal production, or for equipment used in producing these animals. In general, any chemicals which contain phenol or phenol derivatives fall into this category and should not be used in these processes. There are, however, a number of chemical classes which are suitable for these purposes, and the general categories, and characteristics of these are discussed in the following paragraphs. The disinfectants also must not cause harm to personnel using them, or create any stress or damage to the animals being produced.

A. CHLORINE AND CHLORINE-CONTAINING COMPOUNDS

When these compounds are used properly, they are among the most useful, and least damaging, disinfectants available. The chemicals can be in the form of liquid hypochlorite, or can be combined with detergents in crystalline form. Chlorine compounds act rapidly against a wide variety of microorganisms, and are relatively inexpensive. Chlorine compounds are corrosive and have a bleaching action. They therefore must be used carefully around metals, and around materials which may be bleached or faded and thereby harmed. Perhaps the greatest drawback or deficiency to the use of chlorine compounds as disinfectants is that they very readily react with, and are inactivated by, organic compounds other than the target microorganisms. Chlorine compounds, because of this reactivity, should be used in concentrations of at least 100 to 150 ppm (parts per million), and should be allowed sufficient reaction time for killing microorganisms. Since 1988, for use in equipment cleaning, the Food Safety and Inspection Service (FSIS) of the USDA has required the spraying of slaughter equipment for poultry with a solution containing 20 ppm chlorine.[25] This concentration is effective on equipment which may not have a heavy concentration of organic material, and is small enough that no taste alteration will be produced in the poultry. After use of these disinfectants, surfaces treated should be rinsed with water of good quality which is not contaminated.

B. IODOPHORS

Iodine-containing compounds, or iodophors, are similar in many characteristics to chlorine compounds. That is, they are corrosive to metals, readily inactivated by organic chemicals, have a rapid action, and are effective against a wide range of microorganisms. Iodophors are always blended with detergents in an acid medium, and are therefore particularly effective where an acid cleaner is required. These disinfectants should be used at a concentration of 25 to 50 ppm active iodine, and again, allowed sufficient time for action. These compounds are not toxic when used in proper concentrations, but will combine with substances in foods to cause a change in flavor, and, therefore, surfaces and equipment should be thoroughly rinsed after these disinfectants are used.

C. QUATERNARY AMMONIUM COMPOUNDS

Quaternary ammonium compounds are detergents, as well as disinfectants. They have good cleaning characteristics, and are colorless and essentially noncorrosive. Although these chemicals are nontoxic, they do have a bitter taste, and as such are likely to alter the flavor of foods. Quaternary ammonium compounds are not as effective against the common gram-negative bacteria which are members of the Enterobacteriaceae as are chlorine and iodine compounds. Solutions of these chemicals tend to adhere to surfaces, and therefore, have some residual action, but because of such adherence, the rinse process must be more effective to avoid contamination of foods. Because of reactivity with magnesium and calcium compounds, these chemicals can not be used in hard water, and are incompatible with soaps and many detergents.

D. OTHER CHEMICAL DISINFECTANTS

Other chemicals which can be used for special purposes in disinfection are amphoteric surfactants, strong acids, and strong alkalis. The amphoteric surfactants have good detergent as well as killing properties, and are relatively nontoxic. These chemicals do have strong reactivity to organic materials, and are therefore not better than the chlorine and iodine compounds in this respect. Strong acids and alkalis, while having strong antimicrobial activity, are generally corrosive and toxic, and must be used carefully and only in specialized circumstances. Special care must be taken with these materials to assure that foods are not contaminated.

When any disinfectant is selected for use in a certain situation, it must always be monitored to assure effectiveness. In the use of any disinfectant, the absolutely essential considerations must always be the proper killing concentration of the disinfectant matched with the proper application time for the chemical. These two factors must be correlated in order to obtain desired results with any disinfection process. In addition, as mentioned in several places above, these two factors must also be combined with the proper temperature to attain desired results. Some disinfectant chemicals simply are not active, or are not as active if temperature is varied to any great extent from the optimum. The use of chemical disinfectants around areas or equipment used in food processing or preparation is strictly regulated by many agencies. Many chemicals are considered to be safe when used properly, and the specific characteristics of these chemicals are helpful in deciding which should be used in specific situations. Such a characterization is shown in Table 4.1.

The use of sanitizing and disinfecting chemicals for cleaning is closely regulated by the Food and Drug Administration when these materials are used in food processing or food handling operations. The regulations and limitations affecting these chemicals were spelled out in great detail in the Code of Federal Regulations, 1977 and are altered as new types of compounds are brought into use.[26] Among the types of limitations specified are that the agents be used only if permitted by prior sanction and approval, and that the chemicals be used prior to equipment contact with food being processed. Some of these chemicals which are considered safe for use are listed in Table 4.2. Thus the wide variety of chemicals allowed for use, to which newer compounds are frequently added; however, each one, or each combination may be used only if conforming to the proper chemical formulation, and in the proper dilution at the proper stage of food processing. Each chemical has a maximum concentration specified, and this maximum may not be exceeded in any use. If chemicals are used in accordance with these regulations they will bear the proper labeling meeting the requirements of the Federal Insecticide, Fungicide, and Rodenticide Act.[8]

An indication of the fact that the natural habitat of the salmonellas is the intestinal tract is the fact that these organisms were first detected and isolated from clinical specimens, primarily feces.[27] Methods used for isolation of the organisms from animal or clinical specimens have been modified to better fit the behavior of the organisms which are found contaminating foods. The great majority of salmonella are of a low host specificity and exist, even without causing disease symptoms in all cases, in a wide variety of animals.[28,29]

To control the spread of salmonellas, it is essential that all forms of environmental sanitary programs be regularly and rigidly carried out.[8] Since the organisms occupy natural habitats in the intestinal tracts of man and most other animal species, they will be found in a great many locations at all times,

TABLE 4.1
Types, Functions, and Limitations of Cleaning Agents Used in the Food Industries

Categories of aqueous cleaners	Approximate concentrations for use (%)[b]	Examples of chemicals used[c]	Functions	Limitations
Clean water	100	Usually contains dissolved air and soluble minerals in small amounts	Solvent and carrier for soils, as well as chemical cleaners	Hard water leaves deposits on surfaces; Residual moisture may allow microbial growth on washed surfaces; Promotes rusting of iron
Strong alkalis	1–5	Sodium hydroxide; Sodium orthosilicate; Sodium sesquisilicate	Detergents for fat and protein; Precipitate water hardness; Produce alkaline pH	Highly corrosive; Difficult to remove by rinsing; Irritating to skin and mucous membranes
Mild alkalis	1–10	Sodium carbonate; Sodium sesquicarbonate; Trisodium phosphate; Sodium tetraborate	Detergents; Buffers at pH 8.4 or above; Water softeners	Mildly corrosive; High concentrations are irritating to skin
Inorganic acids	0.5	Hydrochloric; Sulfuric; Nitric; Phosphoric; Sulfamic	Produce pH 2.5 or below; Remove precipitates from surfaces	Very corrosive to metals, but can be partially inhibited by amines; Irritating to skin and mucous membranes
Organic acids	0.1–2	Acetic; Hydroxyacetic; Lactic; Gluconic	Remove inorganic precipitates and other acid-soluble substances from surfaces	Moderately corrosive, but can be inhibited by various organic nitrogen compounds

		Citric Tartaric Levulinic Saccharic		
Anionic wetting agents	0.15 or less	Soaps Sulfated alcohols Sulfated hydrocarbons Aryl-alkyl polyether sulfates Sulfonated amides Alkyl-aryl sulfonates	Wet surfaces Penetrate crevices and woven fabrics Effective detergents Emulsifiers for oils, fats, waxes, and pigments Compatible with acid or alkaline cleaners and may be synergistic	Some foam excessively Not compatible with cationic wetting agents
Nonionic wetting agents	0.15 or less	Polyethenoxyethers Ethylene oxide-fatty acid condensates Amine-fatty acid condensates	Excellent detergents for oil Used in mixtures of wetting agents to control foam	May be sensitive to acids
Cationic wetting agents	0.15 or less	Quaternary ammonium compounds	Some wetting effect Antibacterial action	Not compatible with anionic wetting agents Inactivated by many minerals and some other soils
Sequestering agents	Variable (depending on hardness of water)	Tetrasodium pyrophosphate Sodium tripolyphosphate Sodium hexametaphosphate Sodium tetraphosphate Sodium acid pyrophosphate Ethylenediaminetetraacetic acid (sodium salt) Sodium gluconate with or without 3% sodium hydroxide	Form soluble complexes with metal ions such as calcium, magnesium, and iron to prevent film formation on equipment and utensils See also strong and mild alkalis above	Phosphates are inactivated by protracted exposure to heat Phosphates are unstable in acid solutions

TABLE 4.1 (continued)
Types, Functions, and Limitations of Cleaning Agents Used in the Food Industries

Abrasives	Variable	Volcanic ash Seismotite Pumice Feldspar Silica flour Steel wool Metal or plastic "chore balls" Scrub brushes	Removal of dirt from surfaces with scrubbing Can be used with detergents for difficult cleaning jobs	Scratch surfaces Particles may become imbedded in equipment and later appear in food Damage skin of workers
Chlorinated compounds	1	Dichlorocyanuric acid Trichlorocyanuric acid Dichlorohydantoin	Used with alkaline cleaners to increase peptizing of proteins and minimize milk stone deposits	Not germicidal because of high pH Concentrations vary depending on the alkaline cleaner and conditions of use
Amphoterics	1–2	Mixtures of a cationic amine salt or a quaternary ammonium compound with an anionic carboxy compound, a sulfate ester, or a sulfonic acid	Loosen and soften charred food residues on ovens or other metal and ceramic surfaces	Not suitable for use on food contact surfaces
Enzymes	0.3–1	Proteolytic enzymes produced in cultures of aerobic, spore-forming bacteria	Digest proteins, and other complex organic soils	Inactivated by heat Some people become hypersensitive to the commercial preparations Some have contained *Salmonella*

[a]Based largely on information assembled by Elliott (1980).

[b]Concentration of cleaning agent in solution as applied to equipment.

[c]Some regulatory agencies require prior approval (e.g., see U.S. Department of Agriculture, 1977).

From International Commission on Microbiological Specifications for Foods, *Microbial Ecology of Foods*, Vol. 1, Academic Press, New York, 1980, chap. 14. With permission.

TABLE 4.2
Chemicals Recognized as Safe when Used Prior to Equipment Contact with Food Being Processed

Chemicals in aqueous solutions

Potassium hypochlorite
Sodium hypochlorite
Calcium hypochlorite
Dichloroisocyanuric acid (or sodium or potassium salt)
Trichloroisocyanuric acid (or sodium or potassium salt)
Potassium iodide
Sodium p-toluenesulfonyl chloroamide
Sodium lauryl sulfate
Iodine
Butoxy monoether of mixed (ethylene prooxylene) polyalkylene glycol
Ethylene glycol monobutyl ether
Diethylene glycol monoethyl ether
Elemental iodine
Hydriodic acid
Alpha-(paranenylphenyl)-omega-hydroxypoly (polyethylene)
Polyoxyethylene-polyoxypropylene block polymers
Isopropyl alcohol
Sodium iodide
Sodium dioctylsulfosuccinate
Polyoxyethylene-polyoxlypropylene block polymers
Dodecylbenzenesulfonic acid
Polyoxyethylene-polyoxypropylene block polymers
Butoxy monoether
Alpha-lauryl-omega-hydroxypoly(oxyethylene)
N-alkyl(C12-C15)benzyldimethylammonium chloride
Isopropyl alcohol
Trichloromelamine
Sodium lauryl sulfate
Dodecylbenzenesulfonic acid
N-alkyl benzyl dimethyl ammonium chloride
Sulfonated oleic acid (sodium salt)
Polyoxyethylene-polyoxypropylene block polymers
Alkyl monoether
Butoxy monoether
Lithium hypochlorite
N-alkyl benzyl dimethyl ammonium chloride
N-alkyl dimethyl ethylbenzyl ammonium chloride
Di-n-alkyl benzyldimethylammonium chloride
Isopropyl alcohol
N-alkyl benzyldimethylammonium chloride
Sodium metaborate
Alpha-terpineol
Sodium dichloroisocyanurate
Tetra-sodium ethylenediaminetetraacetate

TABLE 4.2 (continued)
Chemicals Recognized as Safe when Used Prior to Equipment
Contact with Food Being Processed

Chemicals in aqueous solutions

Ortho-phenylphenol
Ortho-benzyl-para-chlorophenol
Para-tertiaryamylphenol
Sodium-alpha-alkyl-omega-hydroxypoly(oxylene) sulfate
Dodecylbenzenesulfonate

From U. S. Code of Federal Regulations, 1977.

whether growing or simply remaining viable. These programs must include the sanitary disposal of human and animal waste; the control of pests of all types, particularly rodents and insects; the treatment of environments known to be or likely to be contaminated with pathogenic organisms; and careful control of all sources of foods and waters which are to be consumed by the human.

REFERENCES

1. **Buchanan, R. E. and Gibbons, N. E., Eds.,** *Bergey's Manual of Determinative Bacteriology,* Williams & Wilkins, Baltimore, 1974.
2. **Kreig, N. R. and Holt, J. G., Eds.,** *Bergey's Manual of Systematic Bacteriology,* Vol. 1, 1st ed., Williams & Wilkins, Baltimore, 1984.
3. **Frobisher, M., Hinsdill, R. D., Crabtree, K. T., and Goodheart, C.R.,** *Fundamentals of Microbiology,* 9th ed., W. B. Saunders, Philadelphia, 1974.
4. **Burrows, W.,** *Burrows Textbook of Microbiology,* 22nd ed., Revised by Bob A. Freeman, W. B. Saunders, Philadelphia, 1985.
5. **Altekruse, S. F.,** The Transmission and Spread of *Salmonella enteritidis,* Presentation at the Animal and Plant Inspection Service SE Workshop, Kansas City, MO, June 1990.
6. **Haddock, R. L., Duguies, L. A., and Malilay, J.,** Salmonellosis on Guam: A Problem of Contaminated Environment, in *Proc. Int. Symp. Salmonella,* Snoyenbos, G. H., Ed., American Association of Avian Pathologists, New Bolton Center, Guam, 1985.
7. **Atlas, R. M. and Barta, R.,** *Microbial Ecology: Fundamentals and Applications,* Addison-Wesley, Menlo Park, CA, 1981, 171.
8. **Guthrie, R. K.,** *Food Sanitation,* 3rd ed., Van Nostrand Reinhold, New York, 1988.
9. **Thain, J. A. and Cullen, G. A.,** Detection of *Salmonella typhimurium* in chickens, *Vet. Rec.,* 102, 143, 1978.
10. **Thain, J. A.,** An evaluation of the microantiglobulin test monitoring experimental *Salmonella* group C infections in chickens, *Res. Vet. Sci.,* 28, 212, 1980.

11. **WHO/WAVFH Round Table Conference,** The Present Status of the Salmonella Problem (Prevention and Control), Document VPH/81.27 World Health Organization, Bilthomen, The Netherlands, 1980.
12. **Kaufman, A. F.,** Pets and Salmonella infection, *J. Am. Vet. Med. Assoc.,* 149, 1655, 1966.
13. **Edel, W., van Schothorst, M., Guinee, P. A. M., and Kampelmacher, E. II.,** Mechanism and prevention of *Salmonella* infections in animals, in *The Microbiological Safety of Foods,* Hobbs, B. C., and Christian, J. H., Eds., Academic Press, New York, 1972.
14. **Williams, B. M.,** Environmental Considerations in Salmonellosis, *Vet. Rec.,* 96, 318, 975.
15. **Schor, D.,** The FSIS Challenge — Keeping Pace with New Products, New Microbes, *Food News for Consumers,* Winter, 8, 1990.
16. **Snoyenbos, G. H., Weinack, O. M., Soerjadi-Liem, A. S., Miller, B. M., Woodward, D. E., and Weston, C. R.,** Large scale trials to study competitive exclusion of *Salmonella* in chickens, *Avian Dis.,* 29, 1004, 1985.
17. **Food and Drug Administration,** Human Foods. Current Good Manufacturing Practice (Sanitation) in Manufacture, Processing, Packing, or Holding, Part 128, 21CFR, U.S. Government Printing Office, Washington, D.C., 1974.
18. **DuPont, H. L. and Steele, J. H.,** Use of antimicrobial agents in animal feeds: implications for human health, *Rev. Infect. Dis.,* 9(3), 447, 1987.
19. **International Commission on Microbiological Specifications for Foods,** *Microbial Ecology of Foods,* Vol. 1, Academic Press, New York, 1980.
20. **Craun, G. F., Ed.,** *Waterborne Diseases in the United States,* CRC Press, Boca Raton, FL, 1971.
21. **Benenson, A. S., Ed.,** *Control of Communicable Diseases in Man,* 14th ed., American Public Health Association, Washington, D.C., 1985.
22. **Williams, J. E. and Wittemore, A. D.,** Microtesting for Avian Salmonellosis, in Proc. 77th Meet. U.S. Animal Health Assoc., Washington, D.C., 1973, 607.
23. **Williams, J. E. and Wittemore, A. D.,** Serological response of chickens to *Salmonella thompson* and *Salmonella pullorum* infections, *J. Clin. Microbiol.,* 9, 108, 1979.
24. **Codex Alimentarius Commission,** Cleaning and disinfection, Appendix I, in *Recommended International Code of Practice General Principles of Food Hygiene,* CAC/RCP 1-1969, 1st revision, U.S. Government Printing Office, Washington, D.C., 1979.
25. **Swacina, L.,** Putting Salmonella on hold, *Food News for the Consumer,* 5(3), 12, 1988.
26. Sanitizing Solutions, 21 Code of Federal Regulations, 178.1010, U.S. Department of Health Education and Welfare, Washington, D.C., 1977.
27. *FDA Bacteriological Analytical Manual,* 6th ed., Food and Drug Administration, Association of Official Analytical Chemists, Arlington, VA, 1984, chap. 7.
28. **Stanier, R. Y., Dudoroff, M., and Adelberg, E. A.,** *The Microbial World,* 3rd ed., Prentice Hall, Englewood Cliffs, NJ, 1970.
29. **Edmonds, P.,** *Microbiology: An Environmental Perspective,* Macmillan Publishing, New York, 1978.

Chapter 5

ANTIBIOTIC SENSITIVITY OF *SALMONELLA*

The *Salmonella*, like most other bacteria, are susceptible to a wide variety of chemicals which can be used to inhibit or to control the growth of the bacteria. The chemicals which affect the growth of these organisms are not just those compounds frequently used as chemotherapeutics, but include some of the natural constituents of such foods as onions, garlic, spices, and cocoa.[1] These bacteria are variably sensitive to the usual disinfectants, antibiotics, and other chemotherapeutics. A strain of *Salmonella typhi* is one of the two reference bacteria approved and routinely used in the determination of Phenol Coefficients of disinfectant chemicals.[2]

Disinfectant chemicals may be tested in a manner to permit comparison of effectiveness using the Phenol Coefficient. This is a determination of the amount of dilution of a disinfectant which can be permitted and which will leave the disinfectant in question equivalent in effectiveness of action. The protocol for this test is as follows:

1. Use either *Salmonella typhi,* American Type Culture Collection (ATCC) #6539, or *Staphylococcus aureus* ATCC #6538 and prepare these cultures for the test according to the methods of the Association of Official Analytical Chemists.[3]
2. Compare the effectiveness of chemicals to phenol on the basis of concentration (dilution) and time as given in the Table 5.1.
3. If chemical (X) allows growth in a 1:150 dilution at 5-min exposure, and if phenol allows growth in 1:90 dilution at 5-min but not 10-min exposure, then the phenol coefficient is 150/90 = 1.66, and chemical (X) can be used with confidence at a dilution of 1.66 greater than the dilution required for killing by phenol.

When one considers the reactions of bacteria to chemical compounds, the organisms must be considered to be either sensitive (susceptible) or resistant to the agent. Generally, one considers bacteria to be resistant to a chemical if that organism does not show susceptibility which may be expected. For example, one considers that most microorganisms are sensitive or susceptible to phenol, and if one strain is not inhibited by rather high concentrations of this chemical, we would consider it to be phenol resistant.

While the susceptibility of salmonella to disinfectant chemicals can be considered in the manner applied when chemicals are compared to phenol, an entirely different class of chemical, and entirely different reaction to chemicals is observed in the response of bacteria, including salmonella to chemotherapeutic or antibiotic agents. The purposes of using the chemotherapeutic or antibiotic

TABLE 5.1
Reading the Phenol Coefficient

Concentration (dilution)	Time of Exposure (min)		
	5	10	15
Chemical (X)			
1:50	No growth	No growth	No growth
1:100	No growth	No growth	No growth
1:150	Growth	No growth	No growth
1:200	Growth	Growth	No growth
1:250	Growth	Growth	Growth
1:300	Growth	Growth	Growth
Phenol			
1:70	No growth	No growth	No growth
1:80	No growth	No growth	No growth
1:90	Growth	No growth	No growth
1:100	Growth	Growth	No growth
1:110	Growth	Growth	Growth
1:120	Growth	Growth	Growth

agents is for the inhibition of growth of bacteria which are causing infection in the human or other animals. All chemotherapeutics, if recognized as such, at some concentration will have inhibitory action against some bacterium. Some are totally ineffective in inhibition of certain bacteria and are highly inhibitory to others. The sensitivity or resistance of the bacteria is usually measured *in vitro* in cultures taken from patients or the environment, and such reactivity may not actually correlate exactly with the reactivity of the organism in the environment from which it was taken. The observed sensitivity or reactivity of the organism in a patient, for example, may not be the same as that which will be observed when the organism is cultured from that patient. The use of antibiotics or chemotherapeutics in patients to treat infections may lead to the modification of the state of sensitivity of the organism. Some authors state that treatment of uncomplicated enterocolitis caused by *Salmonella* may lead to the emergence of resistant strains of bacteria[4,5] and such treatment is therefore not recommended.

It is recommended that antibiotics be used to treat enteric fever whenever there is bacteremia. When there is fever, vomiting, and diarrhea in nontyphoid salmonellosis, antibiotics should not be used. Antibiotics appear to convert the carrier state or a case of gastroenteritis to systemic disease with bacteremia, to prolong excretion of the pathogen in the convalescent and to enhance development of resistant strains. Such resistance development appears to always develop following use of antibiotics[6] and is mediated through episomal mechanisms. In any *Salmonella* infection it appears that the normal gut bacterial flora plays a protective role since the alteration of that flora with antibiotic therapy seems to increase the risk of developing salmonellosis.[7]

The chemotherapeutic agents most often recommended for treatment of salmonellosis are chloramphenicol and ampicillin. It is generally recommended that if these agents are not effective, the use of trimethoprim + sulfamethoxazole (TMP-SMX) be instituted. TMP-SMX as used clinically is a mixture of one part trimethoprim plus five parts of sulfamethoxazole. These same drugs are effective against *S. typhi* and most of the other serovars of *Salmonella*, although the recommendations for use differ in the different infections. While antibiotic treatment is generally recommended for typhoid, many experts feel that the use of antibiotics in other salmonelloses may simply prolong the period of shedding of organisms in the feces without affecting the outcome of the infection otherwise. For this reason, in otherwise healthy adults, antibiotics are frequently not used to treat salmonellosis, but instead the fluid-electrolyte balance is monitored and kept in check as the preferred treatment.[4]

When antibiotics are used for treatment, the specific mode of action against the pathogen is dependent on the antibiotic in use. Each antibiotic group has specific, and often different modes of action, and these differences are important to the sensitivity of the pathogen. The bacitracins, cephalosporins, cycloserines, penicillins (including ampicillin), ristocetin, and vancomycin all act on pathogens by inhibiting the synthesis of cell walls.[5] This same mode of action is involved when ampicillin is used to treat infection. Since this is different from some of the other drugs, it is important in handling cases which do not respond to particular medications. In some cases ampicillin may have a high affinity for the beta-lactamase enzyme produced by certain bacteria, but is not hydrolyzed by it, and therefore is not destroyed and can exert its effect on the pathogen.

Chloramphenicol, the erythromycins, lincomycins, tetracyclines, aminoglycosides, amikacin, gentamicin, kanamycin, neomycin, netilmicin, streptomycin, and tobramycin function by inhibition of protein synthesis in pathogenic bacteria. This different mode of action of chloramphenicol makes it effective against certain bacterial strains which are not affected by ampicillin.[5] Chloramphenicol attaches to a subunit of the ribosome and interferes with the binding of new amino acids to the peptide chain. This drug is primarily bacteriostatic, and when it is withdrawn, the organisms begin to grow again. Some bacteria become resistant by production of the enzyme chloramphenicol acetyltransferase which destroys the activity of the drug. The enzyme production within the pathogen is controlled by the presence of a plasmid specific for this activity.

TMP-SMX exerts action on the pathogens by inhibition of nucleic acid synthesis, as do nalidixic acid, novobiocin, pyrimethamine, and rifampin. Since trimethoprim and the sulfonamides have the same mode of action, each can be used separately. These drugs block different stages in the sequence of reactions involved in the synthesis of purines, and eventually DNA. Because of this, when used together there is strong enhancement of the effects because of sequential blocking of reactions in this system.[5]

In consideration of the modes of action of the antibiotics as discussed above

TABLE 5.2
Mode of Action — Selected Antimicrobials

    I.   Inhibit cell wall synthesis
            Penicillins, including
                Ampicillin
                Carbinicillin
                Piperacillin
    II. Inhibit nucleic acid synthesis
            Trimethoprim
            Sulfamethoxazole
            Rifampin
    III. Inhibit protein synthesis
            Chloramphenicol
            Tetracycline
            Kanamycin
            Erythromycin
            Clindamycin
    IV. Competitive inhibition — Inhibit PABA[a] use
            Sulfonamides, including
                Triple sulfa
                Sulfamethoxazole
                Sulfapyradine
                Sulfisoxazole

    [a] PABA = para-amino-benzoic-acid.

and as shown in Table 5.2, it is obvious that organisms which are not growing, or reproducing and are not synthesizing structures vital to survival of the cells are not susceptible to the action of drugs. It is equally obvious that microorganisms producing an infection are growing, reproducing, and producing cell structures necessary for the survival of the pathogens. Therefore, when bacteria are in process of producing infection, they are most susceptible to the action of therapeutic agents.

There are five different mechanisms by which microorganisms may become or exhibit resistance to drugs. These are

1.   Produce enzymes which destroy the drug.
2.   The organism may have no permeability or alter their permeability to the drug. Generally speaking, the drug must enter the organism before it can act, just as a pathogen must enter the host before it can initiate an infection.
3.   The microorganisms may lack or alter a structural target (i.e., an essential protein) for the drug.
4.   The pathogen may develop an alternate pathway for reactions which are inhibited by the drug, or the pathways which are present may be so different as to be unaffected.[5] Certain organisms are also capable of

developing or using different enzymes for catalysis of reactions, which are not affected or are less affected by the drug.

Certain conditions producing metabolic inactivity in microorganisms may render them resistant to all therapeutic agents. Some organisms may become essentially metabolically inactive in specific conditions even though they remain viable in the tissues of the host. This happens in the case of Mycobacteria, in some conditions, during these infections; and during such times, the organisms are unaffected by the presence of any drugs applied. A similar condition may occur in the case of organisms which are treated with penicillin, if the bacteria revert to L forms which lack cell walls. While the cells reproduce without the need for cell wall structure, the penicillin drugs are ineffective regardless of concentration of application. These types of resistance are nongenetic and may be termed naturally occurring resistance. Other naturally occurring resistance will occur in organisms which do not use the process inhibited by the presence of the drug. In that case, there is simply no target structure for the drug, and there is no effect on the pathogen.

Once microorganisms are tested and are determined to be either sensitive or resistant to certain chemotherapeutic or antibiotic drugs, subsequent testing may determine that the organisms have developed resistance or have changed resistance to some of the drugs. When this occurs, there generally has been some genetic change in the organisms. This may occur in several ways including chromosomal resistance which occurs as a result of a spontaneous mutation at a site or locus which controls the susceptibility of that organism to that drug. If this occurs and the drug is present in the environment, then the result is the selection of organisms which are resistant, and the elimination of those which are sensitive to the drug. Spontaneous mutations are not thought to occur frequently, and, therefore, are not considered to be the most common cause of development of drug resistance in clinical usage. The occurrence of such mutations is much higher in the case of some antibiotics than in others and must be considered as a possibility in the use of drugs in certain infections (for example, using rifampin or streptomycin in treatment of mycobacterial infections).

To determine the sensitivity or resistance of a pathogen to an antimicrobial drug, as well as the degree of sensitivity or resistance, tests must be done *in vitro*. Sensitivity testing is used often in clinical practice, and it is a useful tool in attempting to establish the source or pathway of transmission of an outbreak of disease from contaminated foods. These tests are affected by a number of factors in the laboratory environment, including

1. The pH of the culture medium
2. The chemical composition of the medium
3. The solubility and the stability of the drug
4. The number of pathogen cells inoculated into the test

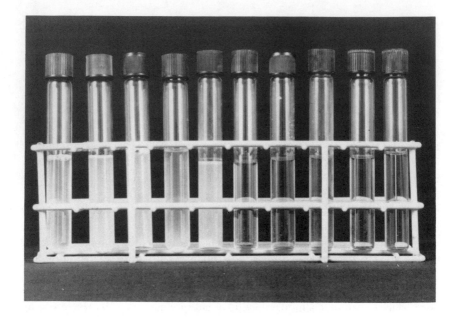

FIGURE 5.1   The tube sensitivity test allows more exact measurement of drug concentration, but is much more expensive and time-consuming to perform.

5.   The length of time the drug is allowed to act
6.   The metabolic state or condition of the organisms

A pure culture of a pathogen in such a test may give a very different response to that observed when the pathogen is included in a mixture of organisms for test. The interpretation of the results of sensitivity tests must take into account such differences, and the preference of the person controlling the test will determine which test is preferred. It is certainly obvious that the pathogen is likely to be present in the infection, or on the contaminated food, in a mixture of populations rather than in pure culture. However, one is interested in the sensitivity of the pathogen, not other contaminating organisms present in the environment, so a decision must be made as to the type of test to be done.[5] When the pathogen is present in the host or the food, it is also subject to the effect of other chemicals in the environment, and these chemicals will affect both the action of the drug and the ability of the drug to reach the pathogen. For example, the proteins present in the host may adsorb much of a drug as it is applied, and the effective concentration of the drug must be in excess of that neutralized by the protein.

Antimicrobial testing is done in one of two ways (Figures 5.1 and 5.2). It may be done by dilution of the drug in a liquid medium, and tested in that medium, or transferred to a solid bacteriologic culture medium from the liquid. The microorganism is then inoculated into the liquid or onto the solid and

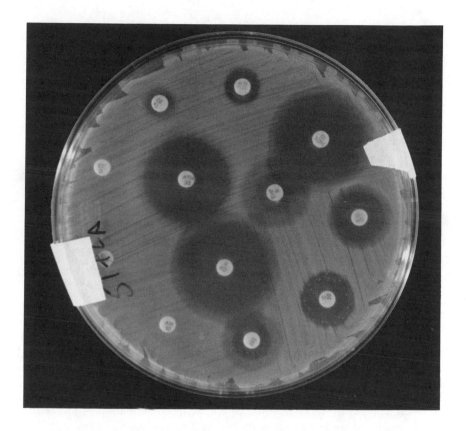

FIGURE 5.2 The disk or plate sensitivity is more rapidly set up than the tube; however, results are somewhat less exact since the size of the zone of inhibition is affected by the solubility and diffusion rate of the antibiotics, by the depth of the culture medium, by the nature of the culture medium, and by the evenness of the surface of the culture medium.

incubated to determine whether growth occurs in the presence of the drug, and, if so, at what concentration of drug does growth occur. This method allows for a more exact measure of the concentration of the drug present at the point of contact with the organism. The second method involves the use of a solid medium for bacterial culture which is evenly seeded or inoculated with a culture of the pathogen. The drug is then applied in the form of different concentrations which have been inoculated into filter paper discs, porous cups, or cylinders which will allow the drug to diffuse. The drug diffuses through the medium, and after incubation, zones around the drug deposits will indicate how much growth inhibition has taken place. In some laboratories, the size of the inhibition zone is measured, and this measurement is taken to indicate the degree of sensitivity or resistance of the organism. Both methods have some advantages, as well as some disadvantages, and these must all be weighed by the investigator to determine the kind of information desired and what will be gained by the method in use.

Most antimicrobial resistance occurs as a result of extrachromosomal factor transfer. The *Salmonella* and many other bacteria are known to contain extrachromosomal genetic factors called plasmids. Some plasmids are called R factors because they are known to carry genes for resistance to one or more antibiotics. Plasmids, or R factors, can be transferred from one bacterial cell to another by the mechanisms discussed in Chapter 2 dealing with the genetic determination of types or serovars within the *Salmonella*. Those mechanisms include transduction, transformation, conjugation, and transposition (see Chapter 2). The only differences involved here and in that discussion are that the genes in this situation control the presence, activity, or specificity of enzymes, pathways, and synthesis of certain structures associated with the drug activity, whereas in the cases discussed earlier, the control was involved with cellular and strain characteristics or specificity of proteins.

The existence of enterobacterial gene transfer systems such as the R factors, is believed to increase the accessibility of the microbial gene pool in any environment, particularly when related species are present.[8] Currently available techniques permit the analysis of R factor characteristics in any strain, and such analysis was done for the strain isolated in the largest outbreak of salmonellosis reported in the U.S.[8] That strain was found to be resistant to many antibiotics, including tetracycline, erythromycin, clindamycin, sulfisoxazole, sulfadiazene, triple sulfa, cefoperazone, streptomycin, mezlocillin, piperacillin, carbinicillin, penicillin, ampicillin, and kanamycin. The drug resistance in this strain distinguishes it from all strains isolated in the U.S. prior to 1984 and is one more link in the chain of evidence indicating that the percentage of salmonellosis outbreaks caused by types resistant to one or more antibiotics is increasing and is making treatment of cases caused by these organisms more difficult. It has been reported that between a study done in 1979 to 1980 and one done in 1984 to 1985, the percentage of resistant strains involved in such outbreaks increased from 16 to 24%.[9]

When two drugs have a common mode of action or related modes of action involving the same or related enzyme sequences, there may occur in microorganisms a condition referred to as cross-resistance to antimicrobials. This situation is relatively common and must be considered in the clinical use of drugs for treatment of infections. It is important in the use of antimicrobials that only those practices which will help to minimize the development of antimicrobial resistance be used. This may be done by maintenance of sufficiently high concentrations of the drugs in animal tissues during use that the original populations of the pathogens, and the emerging first step mutant populations of the organisms are kept under control and not allowed to grow. It is also possible to use drugs in combination in clinical practice so that different mechanisms of action are in effect, and the chances of drug cross-resistance development is reduced.

In 1983, the American Council on Science and Health issued the second edition of their report dealing with the effects of antibiotics in animal feed.[10]

In that report, it was acknowledged that the percentage of bacteria which are resistant to antibiotics does increase when these drugs are used in hospital populations, and that this increase may be related to indiscriminate use of these drugs such as occurred after World War II when antibiotic treatment first became widespread in this country. That report, however, questions the significance of this increase, and the threat posed by such an increase on human health.

Continually, additional strains of *Salmonella* which are resistant to a variety of antibiotics are isolated either from the human or from other animals. The inclusion of subtherapeutic doses of antimicrobial agents in animal feed is often associated with the development of resistance in enteric bacterial flora in these animals. These resistant bacterial forms, without doubt, contribute to the reservoir of resistant bacteria found in the human intestinal tract, including resistant *Escherichia coli* and *Salmonella*. The extent of the use of antimicrobial agents in animal feed is realized when one notes that approximately 45% of the agents used in the U.S. is used in animal feeds. It has been estimated that close to 80% of poultry, 75% of swine, 60% of cattle primed in feedlots, and 75% of calves raised in the U.S. have been fed some antimicrobial agent or antibiotic at some time during life before slaughter and use for food.[11] There is no question that the therapeutic use of these antimicrobials contributes greatly to the development of resistant bacterial strains, and the transmission of these strains to the human. There can also be no doubt that plasmids affecting resistance or sensitivity to these compounds are readily transmitted by many pathways between animals and the human, so that such resistant strains do affect the human populations and do contribute to public health problems. The populations consuming the meats from these animals certainly benefit from the additional growth and lower cost of the animal foods produced in this manner; yet these same populations must also face the consequences of the public health problems initiated by the presence of resistant bacterial populations in the environment. It appears at present that any reduction or control in the use of antimicrobial agents in sublethal doses will assist in the reduction of development of resistant strains of pathogenic organisms. The variety of strains and the different degrees of resistance reported in recent years among the *Salmonella* include a broad spectrum of organisms and a wide divergence of sensitivity of the organisms to many antimicrobial agents. Strains of *S. typhimurium, S. newport, S. krefeld, S. heidelberg, S. saint-paul, S. enteritidis*, and *S. dublin*, among others have been reported as resistant to a variety of antibiotics or chemotherapeutics, including chloramphenicol (frequently chosen as the first drug of choice for treating recalcitrant cases of salmonella infection), chlortetracycline, oxytetracycline, novobiocin, kanamycin, ampicillin, sulfadiazine, streptomycin, TMP-SMX, and tetracycline, among other drugs by many investigators.[12-20]

Gaining consideration in this country is the practice of reducing use or even not allowing the addition of certain antibiotics or chemotherapeutics to animal feeds, or in some instances for any animal use. This restriction is practiced in some countries and is being recommended in others. This prohibition helps to restrict the contact of the microorganisms with the drugs which are most used and are most likely to be needed for infection treatment in the human. In the salmonella, strains carried by animals have developed resistance to antibiotics which were allowed as additives to feeds. The development of resistance in these to the tetracyclines has been frequently observed. Addition of antibiotics to animal feeds has a beneficial effect on the rapidity of development of the animal, but the development of resistance is so detrimental that drug addition to feeds is restricted in Britain.[5]

O'Brien et al.[21] reported on the epidemiology of antibiotic resistance in salmonella and discussed the importance of antibiotic addition to animal feeds from the standpoint of human contact with the organisms. While spread of antibiotic-resistant strains from animals to farm workers has been demonstrated, it has proven much more difficult to prove that any one resistant strain isolated from a human has come from an animal being fed antibiotic-containing feeds. Meats often contain large numbers of enteric bacteria of human origin, and it is not known how important these strains may be in the establishment of populations within the gut of persons not residing on the farm.[11] The frequency of feeding antibiotic-containing diets to animals on the farm has been well documented. Such diets are fed to promote growth, and in addition, there is little control over the use of antibiotics for treatment of disease in farm animals. The use of chloramphenicol is prohibited in animals used for food, although it is licensed for use in pets and other nonfood animals.[22] The transfer of drug-resistant strains of bacteria from animals to the human, as well as the role of addition of antimicrobials to feed for food animals, are still controversial topics; however, such studies as are reported here are beginning to weigh the evidence in favor of controlling the use of such additives. In the California hamburger case reported next, the resistant bacteria did not result from the addition of the drug to feed, but from the use of the drug to treat infections in dairy cattle which were then used for slaughter and for human food.

DuPont and Steele[11] state the case for the control of the use of antimicrobials in animal feeds: "The inclusion of subtherapeutic doses of antimicrobial agents in animal feed is credited for having contributed to lower costs of meat, milk, and eggs. The practice often is associated with the acquisition of resistant enteric flora by the involved animals, a phenomenon that in turn may contribute to the human reservoir of coliforms and salmonellae resistant to antimicrobial agents. Farm workers may transiently acquire resistant intestinal flora and on rare occasions develop salmonellosis. Although irrefutable evidence of the growth-promoting properties of antibiotics in animal feed was provided 30 to 40 years ago, additional studies — with a focus on mechanisms of the effect — are presently needed." Certainly the numbers of antibiotic resistant strains and serovars of salmonellae are increasing, and if addition of antibiotics to feed is one of the causes for that, then strict control is required to at least stop the increase in resistant strains.

Debate continues as to the likelihood that antibiotic resistant strains are more likely to cause infections in the human. DuPont and Steele[11] report that within 1 week of feeding animals certain antibiotic-supplemented foods, most intestinal coliforms become resistant to the introduced antimicrobials. If these bacteria become resistant, there is no evidence that such pathogens as the salmonellae will not do the same. In fact, all the evidence is to the contrary; the pathogens do develop resistance. Epling and Carpenter[23] reported that 84.3% of 121 isolates of *Salmonella* from pork carcasses showed multiple resistance patterns with two or more drugs. The most commonly found resistance was to penicillin, followed by trimethoprim and ampicillin. DuPont and Steele[11] also state that resistant bacteria are found in the flora of farm workers who have close and regular contact with the animals and their antibiotic supplemented feeds on the farm. One study reported by them stated that one of three fecal samples from persons living on a poultry farm where oxytetracycline-supplemented feeds were used, yielded tetracycline-resistant bacteria within 5 to 6 months. This resistance disappeared when feeding of supplemented feed was discontinued. Cohen and Tauxe[17] report that many investigators have concluded that the development of resistance in bacterial strains is related directly to the amount and use of antibiotics in the human and animals. This conclusion leads to a recommendation that these drugs be used more prudently in all situations.

In the U.S., the Food and Drug Administration (FDA) has proposed a limited ban on the use of antibiotics in food animals. In that ban, the use would be limited to therapeutic application of the drugs and would still be available on veterinary prescription. Other countries have tried a similar ban on such use, and in some cases the results have been studied. The U.K. instituted such a ban in 1971, and studies have not found any decrease in the excretion of antibiotic resistant bacteria by farm animals following the ban. On the other hand, studies in the Netherlands have noted a country-wide decrease in tetracycline resistance which coincided with the ban on use of that drug in animal feeds which occurred in 1984. That observed decline occurred in one strain of *S. typhimurium*, and its presence in all other salmonella strains is not known. Studies in this country have indicated that both *Salmonella* and *Campylobacter* are isolated more often from poultry than from other animal products, and that these isolates have continued to show essentially the same resistance patterns to tetracyclines and penicillin which were present before use of these drugs in poultry feeds was discontinued in 1971. The American Council on Science and Health report[10] points out that the evidence on the effect of the use of antibiotics in animal feeds continues to be conflicting, and that no clear solution or recommendation is currently possible.

The Centers for Disease Control (CDC) reported the most commonly isolated serovars of *Salmonella* in the U.S. in the 1987 Annual Summary[24] to be those listed in the Table 5.3. The percentage of total reported isolated strains is listed to illustrate the relative frequency with which one strain might be expected to be encountered. The comparison between those strains isolated from the

TABLE 5.3
Most Frequently Reported Salmonella Serotypes in the U. S., 1987

Serotype	Number reported	Percent of total
Human reports		
Typhimurium	10,462	23.5
Enteritidis	6,950	15.6
Heidelberg	5,714	12.8
Newport	2,858	6.4
Hadar	2,170	4.9
Infantis	1,136	2.5
Agona	1,080	2.4
Montevideo	1,037	2.3
Thompson	635	1.4
Braenderup	548	1.2
		73.1
Total reported	44,609	
Nonhuman reports		
Typhimurium	1,246	13.5
Heidelberg	1,124	12.2
Choleraesuis	667	7.2
Reading	438	4.8
Hadar	352	3.8
Senftenberg	331	3.6
Newport	302	3.3
Montevideo	301	3.3
Enteritidis	270	2.9
Anatum	266	2.9

From *Salmonella* Surveillance, Annual Summary, 1987, Centers for Disease Control, Public Health Service, U.S. Department of Health and Human Services, Atlanta, GA, 1987.

human and those isolated from other sources (primarily other animals) indicates the importance of so-called animal strains to human public health. If one also takes into account the strains reported to have been isolated which were resistant to one or more antimicrobials, it becomes even more apparent that antimicrobial resistance is critically important to human public health. Of the animal isolates which also have been reported to have frequent development of antimicrobial resistance are the top four human isolates (typhimurium, enteritidis, heidelberg, and newport), indicating that any resistance which may develop in these strains, due to the presence of antimicrobials in animal feeds, will have a direct and important bearing on human public health and the control of salmonellosis in the human.

Salmonella typhimurium was listed in 1987[24] as the most frequently isolated serotype in the U.S. This serovar also has a broad host range and causes disease in a wide variety of animal species. Unfortunately, this organism is also frequently reported as being resistant to one or more antimicrobial agents against which it has been tested and multiple-resistant strains are not unusual. It has been reported that 15% of persons who were infected with multiple-resistant strains of this serovar and who were treated with ampicillin, subsequently have ampicillin-resistant strains isolated from the stool.[17] It was suggested that this change in the resistance of the organisms was due to acquisition of resistance factors from other enteric flora. The number of strains of this serovar which are reported to have multiple resistance make up a large portion of the strains reported as isolated from the human. In 1975, Neu et al.[13] reported that this serovar was identified in 34% of the 718 isolates. Cohen and Tauxe[17] reported that this serovar accounted for 35% of the human isolates in 1984. In the 1987 CDC summary,[24] this serovar was found to account for 23.5% of the 44,609 isolates from the human. In 1975, of 718 isolates of *S. typhimurium* from the human, 57.6% were resistant to one or more antibiotics, whereas of 688 isolates from animals, 80% were resistant to one or more antibiotics. In the human isolates, more than half were resistant to four or five antibiotics. Ampicillin resistance in *S. typhimurium* isolates rose from 23 to 37% from 1965 to 1973, whereas resistance to any antibiotic rose from 19 to 58% of isolates.[13] Those authors also reported an unexplained cyclic variation in resistance of this serovar to antibiotics, in which the isolates were more resistant in the winter months than in the spring and summer.

The importance of resistant strains in the course of human infections with salmonella is not yet fully realized. It was reported in 1986 that persons who had previously been exposed to antibiotics were more susceptible to salmonella infection,[17] and that more than 2800 cases of infection occurred in these persons which would not have occurred if the patient had never come in contact with antimicrobials. These investigators also reported that antimicrobials applied to the human appear to convert asymptomatic colonization with salmonella into active cases of infection when the strains of salmonella are antimicrobially resistant, and that the required infectious dose for initiation of disease was lowered in the case of antimicrobially resistant strains. In the 1985 milkborne outbreak of salmonellosis in the Midwest, persons who were infected and were taking antimicrobials to which the organisms were resistant apparently drank significantly less milk than others who were also infected. These results suggest that a smaller inoculum of bacteria was necessary for initiation of the infection in those individuals taking the antimicrobials. The use of antimicrobials for treatment of infection in animals is believed by many to be responsible for at least part of the development of resistance in some strains of salmonella.[17] They point out that an argument in favor of this is the fact that chloramphenicol resistance is increasing in the *S. typhimurium* serovar isolated, and that this serovar is the most commonly isolated in the U.S. where chloramphenicol is infrequently used to treat human infections. This same illustration can also be used to counter the claim that the indiscriminate use of antimicrobials in hospital environments in this country is responsible for development of resistant salmonellas. Antimicrobial resistance in the salmonella corresponds very closely with the use of certain antimicrobial compounds in animals, particularly in animal feeds, but does not correspond closely to the use of other antimicrobials in treatment of the human patients in hospital.

There is certainly no evidence to indicate that *S. typhimurium* is the only serovar, or even the major serovar causing outbreaks of salmonellosis, or that this is the only serovar developing multi-antimicrobial resistance. This serovar has, however, been implicated in some of the major outbreaks of salmonellosis which have been reported in this country. It has been stated in numerous instances that the largest single outbreak of salmonellosis reported in the U.S. was that which resulted from contamination of pasteurized milk in a plant in Illinois in 1985. That outbreak was found to be caused by a strain of *S. typhimurium* which was unusually resistant to a number of antimicrobial compounds.[8,25] In the excellent epidemiological investigation of that epidemic, it was found that the causative organism involved had been responsible for at least three outbreaks of disease in Illinois over a period of 8 months[24] causing as many as 168,000 cases of salmonellosis and resulting in two deaths. Although the largest single outbreak ever recorded in this country and certainly one of the most thoroughly investigated, this milkborne outbreak of salmonellosis is not the only such outbreak recorded in this country. A previous milkborne salmonellosis outbreak (raw instead of pasteurized milk was involved) was reported as being caused by a *S. typhimurium* serovar.[16] That strain was unusual in that it was resistant to chloramphenicol among other drugs, including ampicillin, carbenicillin, kanamycin sulfate, streptomycin, sulfisoxazole, and tetracycline. The epidemiologic investigation of this epidemic determined that almost half of the isolates of this serovar reported to the health authorities in Arizona were resistant to chloramphenicol. Since this drug may be the drug of choice for treating recalcitrant cases of salmonellosis, such resistance greatly impeded the treatment of these infections. In this outbreak there was at least one fatality reported as due to infection by this organism.

Multiple antimicrobial resistance is perhaps most frequently reported in the typhimurium serovar. Unfortunately, this serovar is the one most frequently reported from both human and nonhuman sources in the U.S. Other serovars which have been reported to develop single or multiple resistance to antibiotics are the dublin, gallinarum, choleraesuis, newport, saint-paul, enteritidis, infantis, heidelberg, derby, and san diego serovars. Of these, newport, enteritidis, infantis, and heidelberg are among the ten most frequently reported serovars in the U.S. From nonhuman sources, the choleraesuis, newport, enteritidis and heidelberg serotypes are among the ten most frequently reported. The gallinarum serovar is primarily a poultry pathogen, but has been reported to be capable of infecting the human.

A report in 1987[22] detailed the epidemiology of a chloramphenicol-resistant strain of *S. newport* which was traced through hamburger to dairy farms as the source. This study resulted from an increase in the occurrence of this serovar in California in 1985 and was followed because 87% of the isolates of this strain showed the unusual resistance to chloramphenicol. This report prompted the Texas Department of Health to examine newport isolates in that state, since the serovar ranked as the second most common type in Texas, although in recent years the percentage of newport isolates has declined. When Texas strains were tested, only 5% were found to be chloramphenicol-resistant as compared to the 87% reported in California.[26] Chloramphenicol is the drug of choice for treating recalcitrant cases of salmonellosis; however, it is used only when the necessity is fully recognized in the human because of undesirable side effects in many persons. In the California study, it was discovered that the illness was associated with the use of penicillin or tetracycline during a period prior to the illness, and that there was also an association with the consumption of ground beef the week before the onset of symptoms. When the organisms were found in the meat to be resistant to chloramphenicol, cattle brought in for slaughter were tested. The chloramphenicol-resistant strain was then isolated from cattle on the dairy farms from which the cattle were brought in for slaughter. On three of these farms, chloramphenicol had been used for treatment of infections in the dairy cattle in the previous 18 months. The strain causing the epidemic was found on six of the eight farms, as well as a strain of the dublin serovar with the same antibiotic resistance pattern. Undoubtedly, the use of the drug to treat infection in the dairy cattle had resulted in antibiotic resistance in the strains, which was then transmitted through the contaminated hamburger meat to the consumers.

A different aspect of the outbreak of salmonellosis in California is that hamburger, improperly cooked, was the source of the salmonella. This is best explained by the occurrence of salmonella organisms in beef or dairy cattle. In this epidemic, the contamination had occurred in dairy cattle which were eventually sold for economic reasons for beef. Cattle are the third most frequent source of nonhuman isolations of salmonella as reported in the *1987 Salmonella Surveillance* by the CDC.[24] In that surveillance year, 1653 (approximately 18%) of 9208 isolates from nonhuman sources came from cattle while 52% came from poultry. These numbers were up from the totals of 713 (13%) or 5243 in 1977 when 28% came from poultry and only 3% came from feeds. It was also noted in the 1987 report that 7% of the isolates from nonhuman sources came from animal feeds. This percentage is certainly sufficient to explain the continued spread of salmonella strains in animals, and where antibiotic supplementation of feed is allowed, to help explain continued appearance of antibiotic-resistant strains of these bacteria.

The emergence of antibiotic-resistant enteric pathogens after the use of antimicrobials in the human has been reported a number of times.[11,14,22,25] Obviously, what happens in these instances is that the resistant strain is not able to initiate symptoms of an infection, until competing microorganisms are reduced to allow development of the illness. One factor in such situations is that prior antimicrobial therapy allows fewer numbers of antimicrobial-resistant *Salmonella* to initiate an infection.[17] These writers also state that as the proportions of salmonellas which are resistant to antimicrobials increase, the frequency of salmonellosis will also increase.

The American Council on Science and Health report[10] cites work by Atkinson and Lorian[27] indicating that of 16 commonly used antibiotics in hospitals over a 12-year period, bacterial susceptibility on a national scale remained essentially unchanged. Since antibiotic treatment of uncomplicated salmonellosis other than typhoid is not recommended, the resistance or sensitivity of the causative strain in these patients is relatively unimportant. The importance of little change in resistant strains becomes more apparent in patients with bacteremia, or in the very young or very old because this characteristic of the causative organism may very well determine the outcome of the treatment regime. In a strain, for example, which is resistant to the drugs of choice, or any portion of them, treatment may result in failure or such great delay in effectiveness that the patient suffers unnecessarily. Holmberg et al.[12] have reported that the mortality, likelihood of hospitalization, and the length of hospital stay were usually twice as great when drug-resistant strains of bacteria were involved as when drug-sensitive organisms were the causative organisms. These writers state that this is the case whether the organisms were acquired in the hospital or in the community.

REFERENCES

1. **Litchfield, J. H.,** Salmonella Food Poisoning, in *The Safety of Foods*, 2nd ed., Graham, H. D., Ed., AVI Publishing, Westport, CT, 1980, 120.
2. **Burrows, W.,** *Burrows Textbook of Microbiology*, 22nd ed., Revised by Bob A. Freeman, W. B. Saunders, Philadelphia, 1985.
3. **Association of Official Analytical Chemists,** *Official Methods of Analysis of the Association of Official Analytical Chemists*, Association of Official Analytical Chemists, Washington, D.C., 1980.
4. **Benenson, A. S., Ed.,** *Control of Communicable Diseases in Man*, 14th ed., American Public Health Association, Washington, D.C., 1985.
5. **Jawetz, E., Melnick, J. L., and Adelberg, E. A.,** *Review of Medical Microbiology*, 17th ed., Lange Medical Publications, Los Altos, CA, 1986.
6. **Dupont, H. L. and Hornick, R. B.,** Clinical approach to infectious diarrheas, *Medicine*, 52, 265, 1973.
7. **Larson, E.,** *Clinical Microbiology and Infection Control*, Blackwell Scientific Publications, Boston, 1984.
8. **Schuman, J. D., Zottola, E. A., and Harlander, S. K.,** Preliminary characterization of a food-borne multiple-antibiotic-resistant *Salmonella typhimurium* strain, *Appl. Environ. Microbiol.*, 55, 2344, 1989.

9. **MacDonald, M. L., Cohen, M. L., Hargrett-Bean, N. T., Wells, J. G., Puhr, N. D., Collin, S. F., and Blake, P. A.,** Changes in the antimicrobial resistance of *Salmonella* isolated from humans in the United States, *J. Am. Med. Assoc.*, 258, 1496, 1987.

10. **American Council on Science and Health,** *Antibiotics in Animal Feed: A Threat to Human Health,* 2nd ed., American Council on Science and Health, Summit, NJ, 1985.

11. **Dupont, H. L. and Steele, J. H.,** Use of antimicrobial agents in animal feeds: implications for human health, *Rev. Infect. Dis.*, 9, 447, 1987.

12. **Holmberg, S. D., Solomon, S. L., and Blake, P. A.,** Health and economic impacts of antimicrobial resistance, *Rev. Infect. Dis.*, 9, 1065, 1987.

13. **Neu, H. C., Cherubin, C. E., Longo, E. D., Flouton, B., and Winter, J.,** Antimicrobial resistance and R-factor transfer among isolates of *Salmonella* in the northeastern United States: a comparison of human and animal isolates, *J. Infect. Dis.*, 132, 617, 1975.

14. **Holmberg, S. D., Osterholm, M. T., Senger, K. A., and Cohen, M. L.,** Drug-resistant Salmonella from animals fed antimicrobials, *N Engl. J. Med.*, 311, 617, 1984.

15. **Riley, L. W., Cohen, M. L., Seals, J. E., Blaser, M. J., Birkness, K. A., Hargrett, N. T., Martin, S. M., and Feldman, R. A.,** Importance of host factors in human salmonellosis caused by multiresistant strains of *Salmonella, J. Infect. Dis.*, 149, 878, 1984.

16. **Tacket, C. O., Lee, B. D., Fisher, H. J., and Cohen, M. L.,** An outbreak of multiple-drug-resistant *Salmonella* enteritis from raw milk, *J. Am. Med. Assoc.,* 253, 2058, 1985.

17. **Cohen, M. L. and Tauxe, R. V.,** Drug-resistant *Salmonella* in the United States: an epidemiologic perspective, *Science,* 234, 964, 1986.

18. **Lecos, C. W.,** Of microbes and milk: probing America's worst *Salmonella* outbreak, *FDA Consumer,* U.S. Department of Health and Human Services, Washington, D.C., February 1986, 18.

19. **Mathewson, J. J., Simpson, R. B., and Roush. D. A.,** Isolation of antibiotic-resistant *Salmonella krefeld* from clinical veterinary materials, *Antimicrob. Agents Chemother.*, 19, 355, 1981.

20. **Mathewson, J. J. and Murray, B. E.,** Plasmid-mediated resistance to trimethoprim-sulfamethoxazole in *Salmonella krefeld* strains isolated in the United States, *Antimicrob. Agents Chemother.*, 23, 495, 1983.

21. **O'Brien, T. F., Hopkins, J. D., Gllleece, E. S., Medeiros, A., Kent, R. L., Blackburn, B. O., Holmes, M. B., Reardon, J. P., Vergeront, J. M., Schell, W. L., Christensen, E., Bissett, M. L., and Morse, E. V.,** Molecular epidemiology of antibiotic resistance in Salmonella from animals and human beings in the United States, *N. Engl. J. Med.*, 307, 1, 1982.

22. **Spika, J. S., Waterman, S. H., Soo Hoo, G. W., St. Louis, M. E., Pacer, R. E., James, S. M., Bissett, M. L., Mayer, L. W., Chiu, J. Y., Hall, B., Greene, K., Potter, M. E., Cohen, M. L., and Blake, P. A.,** Chloramphenicol-resistant *Salmonella newport* traced through hamburger to dairy farms, *N. Engl. J. Med.*, 316, 565, 1987.

23. **Epling, L. K. and Carpenter, J. A.,** Antibiotic resistance of *Salmonella* isolated from pork carcasses in northeast Georgia, *J. Food Prot.*, 53, 253, 1990.

24. **Centers for Disease Control,** Annual Summary 1987, *Salmonella Surveillance*, Centers for Disease Control, Public Health Service, U.S. Department of Health and Human Services, Atlanta, GA, 1987.

25. **Ryan, C. C., Nickels, M. K., Hargrett-Bean, N. T., Potter, M. E., Endo, T., Mayer, L., Langkop, C. W., Gibson, C., MacDonald, R. C., Kenney, R. T., Puhr, N. D., McDonnell, P. J., Martin, R. J., Cohen, M. L., and Blake, P. A.,** Massive outbreak of antimicrobial-resistant salmonellosis traced to pasteurized milk, *J. Am. Med. Assoc.,* 258, 3269, 1987.

26. **Texas State Department of Health,** Bureau of Epidemiology, *Texas Preventable Disease News*, 47, 1, 1987.

27. **Atkinson, B. A. and Lorian, V.,** Antimicrobial agent susceptibility patterns of bacteria in hospitals from 1971 to 1982, *J. Clin. Microbiol.*, 20, 791, 1984.

Chapter 6

PREVENTING THE CONTAMINATION OF FOODS BY *SALMONELLA*

The Hazard Analysis Critical Control Point (HACCP) system was first presented at the 1971 National Conference on Food Protection.[1] It consists of three parts: (1) identification and assessment of hazards associated with growing, harvesting, processing, manufacturing, marketing, preparation, and use of a raw material or food product; (2) determination of critical points at which identifiable hazards may be controlled; and (3) establishment of procedures to monitor the identified control points to determine whether or not a hazard does occur. The key to the success of any HACCP program lies in the meaning and recognition of the critical control point. This is the point in a process at which the product may be contaminated with pathogenic microorganisms, food spoilage organisms, or toxic chemicals unless the point and the process are adequately controlled.

There is a constantly increasing demand for prepackaged foods requiring less time-consuming work for the housewife. Thus, the burden is being placed on the food processor to produce a product of such quality that the housewife can rely on the quality to protect the family of consumers. Those who work in the field are in agreement that there is no such thing as absolute purity and no zero-chance of contamination. Unavoidable defects are objectionable as is minute contamination, but realistically these things are going to happen. These types of occurrences are among the reasons that the Food and Drug Administration (FDA) of the U.S. continues to increase the surveillance and regulation of food processing and production although the U.S. has the safest food supply in the world.[2] To ensure that this quality is maintained, the FDA initiated and hopefully will expand the program commonly referred to as HACCP. This system, inaugurated in inspections, follows a necessary system of controls inserted at points where control is critical in the sense that these are the points where food contamination may occur or would result in some unacceptable defect in the final product. In food processing, such points are frequently masked because the product will later be treated by some method (heat or other processing) which will negate any bacterial contamination or cover the taste or odor of chemical contaminants which may have entered at this point. The inspection technique termed HACCP includes three parts: (1) a traditional inspection of the plant covering the processing of a day and inspecting the flow chart of the process being used and identification of the critical points where quality control must be applied in that process, (2) determination of the extent of the processing company's own quality assurance program covering the control points identified, and (3) documentation of the extent to which the company adheres to its own program. This system provides for the FDA a means of estimating the satisfactory operation of the plant for each day whether an inspector is present or not. It also helps to identify potential problems with the product produced and provides more specific corrections when objectionable practices need to be corrected.

HACCP is a careful systematic approach to assure food safety. In this system, a procedure is set up to establish food safety, to monitor, and quickly correct any problems. It is an analysis of procedures to assure food safety in any food handling operation, whether it be production, processing, or service. HACCP, although introduced and recognized for several years, is still primarily applied to food service rather than to other food operations. HACCP has been used for some food processes for over 20 years, but is just recently being used by the Food Safety Inspection Service (FSIS) of the U.S. Department of Agriculture (USDA).[3] Perhaps a part of the reason for this is the fact that the step by step operation of the system has never been agreed to by all parties concerned and that lack of agreement has not permitted the establishment of specific Good Manufacturing Practices (GMPs) which are interpreted to have the force of law. Without doubt, when such regulations are established and accepted as standard practice, we can accomplish much better control of the transmission of foodborne diseases.

Use of HACCP has resulted in the issuance of many GMP regulations which enhance the proper enforcement of Section 402 of the Federal Food, Drug, and Cosmetic Act of 1938 as amended many times since, but are still not sufficiently specific to attain the force of law. That section of the Act is the basis for enforcement of most of the regulations protecting the quality of our food source and states: "A food shall be deemed to be adulterated if it has been prepared, packed, or held under insanitary conditions whereby it may have become contaminated with filth or whereby it may have been rendered injurious to health." GMPs issued by the FDA aid industry by allowing it to more readily recognize what the FDA is looking for in regard to quality assurance. These are required because of the vague language and terms included in many of the laws which have been enacted to regulate the food processing industry.[4] GMPs help avoid the necessity for recalls or other legal actions which do nothing to increase consumer confidence in the quality of foods provided by the industry under the supervision of the FDA. Some GMPs are written with vague language giving the FDA inspector broad latitude in the enforcement of the regulation, and in some cases require interpretation by the inspector in order that the industry may have a better understanding of what is required by the FDA. Such GMPs are referred to as "umbrella" GMPs, and the vague language used has led to some court rulings that these lack the force of law ascribed to the more specific GMPs. For the GMP to have the force of law, specificity must be such that the processor knows exactly what must be done, and the regulator or inspector can accurately determine adherence or lack of compliance with the regulation. Sometimes inspections by the FDA or other health agencies are necessary because certain conditions can only be recognized and properly corrected during plant inspections. Those things that can be recognized by inspections are the easier to correct, but some things are not recognizable by inspection, including specific pathogenic bacterial contamination.

The National Research Council, Food and Nutrition Board, Committee on Food Protection[5] indicated that a major factor in the success of HACCP application to food safety was the fact that this system was mandated by federal regulation of the food industry. The committee concluded that several factors are of much importance for the HACCP system to be broadly applied in the food industry.

1. Technical sophistication must be available for use in the hazard analysis of a food process and in the identification of control points with the establishment of monitoring programs. Experts in the safety of foods continuously point up the essentiality of proper training for the inspectors, regulators, and the food industry personnel operating the HACCP program. It was the recommendation of this committee that this training be accomplished as a function of the industry trade associations. The reason for this recommendation was that if done in this way, all industry companies will participate, and the results will not be detrimentally affected by leaving out the less sophisticated and smaller companies. Failure of these smaller companies to use the HACCP system will expose their products to greater hazard of microbiological contamination and over the long run will cost these companies both financially and in consumer confidence. This latter effect will not be felt just by the smaller companies, but by the industry as a whole. In addition to the training of the industry representatives involved with HACCP, the regulating agencies must also have properly trained personnel, not only to monitor the programs, but to promulgate the specific GMPs which can be assigned the force of law by the courts.

2. The training within the regulatory agencies must not be applied entirely to those who are monitoring the program, but must be expanded to all personnel involved in and responsible for the regulation, initiation, and implementation of the program. Failures to continue and update training in these areas will lead to breakdowns within the program, resulting in hazardous conditions in the food industry and failure to use the system except where it is mandated by law.

3. The success realized in the use of HACCP in controlling the processing of low-acid canned foods has resulted in part because of the mandatory training of technicians to carry out the program. If HACCP is to be successfully used throughout the industry to reduce the occurrence of hazards, then it is necessary that training of individuals in all branches of the food industry be carried out on an efficient and regular basis.

4.   To do this, the issuance of specific GMPs for each segment of the industry will contribute further to the success of application of the HACCP program. The mandate applied to low-acid canned foods following the most specific GMP is believed by most to be a major factor in the success of that program. It is clear from this, that voluntary compliance is not successful with such programs over a long period of time when hazards which cause difficulty for the industry are not encountered.

The major problem associated with the establishment or implementation of a HACCP program for any product in any processing plant is the fact that it is voluntary, and unless the processor sees the obvious benefit, there will be no pressure from the company to install the program. The regulators have not initiated the mandatory programs, and until this is done, it is unlikely that the industry will do so. It has become more common that the HACCP approach has been used by local regulators for controlling food service establishments, and such use has paid dividends in many localities by reducing the points at which foods are contaminated where food is served directly to the consumer. Many critical control points have been recognized and have been detailed for food service establishments by the Food Marketing Institute.[5] In most cases the points identified for the food service programs would work equally well for food processing programs. A HACCP program for any establishment may be set up during a routine inspection and should include several points: (1) review of the process, (2) asking questions to be sure that the complete process is understood, (3) following the food for development of accurate flow charts, (4) determining whether the food is hazardous or non-hazardous and making sure that it meets all criteria for that category, (5) documenting all times and temperatures, (6) thinking cross-contamination throughout the process and looking for possible sites where this may occur, (7) taking samples and pictures during the inspection so that there will be sufficient evidence for later discussions and decisions, and (8) making recommendations and corrections after having adequate time for consideration of the flow of the process.

A hazard analysis is an evaluation of an entire process, including all its steps or procedures during the production, processing, distribution, and use of raw materials or finished food products. It must include identification of potentially hazardous raw materials or food products which may contain injurious substances, whether they be chemical contaminants or microorganisms. It is essential that all components of the food product which might support microbial growth be identified in that product. It is also essential that all sources and specific points in the process where contamination can occur be identified, and that the potential for microorganisms or toxic chemicals to persist during a process be identified, as well as the potential for microorganisms to multiply during the process. It becomes obvious, therefore, that hazards mean any unacceptable growth, survival, or presence of microorganisms or chemicals of concern to the safety or spoilage of raw materials, products, or components of products during processing, distribution, or storage of the product before or during marketing.

During a hazard analysis it is essential that answers be found to all what?, why?, when?, where?, and how much? questions concerning the product, the product formulation, the process, and the conditions of intended distribution and use of the products. When these answers are obtained, only then can there be a preliminary assessment of the potential for hazard and an evaluation of the product safety and stability. The hazard analysis may require, in some cases, that the product be inoculated with certain commonly foodborne pathogenic organisms to test the potential for survival, growth, and maintenance of pathogenic or spoilage organisms. Once inoculated, the food or product must then be subjected to the usual process to detect the effect of different steps on the contaminants, as well as the effect of mishandling on the final product.[6]

Test protocols such as the above help to detect the points or stages of a procedure where hazards exist. Once these are detected, then the control points for elimination can be identified. It is at these critical control points (Figures 6.1 and 6.2) that hazards must be corrected or removed to prevent unacceptable production or persistence of microorganisms and/or their metabolic products. Critical control points are those which are not followed by processes or operations which will eliminate the hazard that has been detected. Time and temperature of any operation are always critical control points. Time-temperature relationships and failures therein are the major defects in most processes of any food product past the raw material.[7] These time-temperature relationships which are critical control points in processing also are equally, if not more, critical in holding, cooling, and freezing periods as well as other operational processes.

Cleaning of equipment that contacts any product during the heat processing, holding, cooling, or freezing periods also constitutes one or more critical control points depending upon the cleaning process used. Other critical control points in any process become apparent during a hazard analysis of a particular process. Sometimes these points are not detected until after an outbreak of illness when similar processes are used. These events are unfortunate in that all outbreaks can not be prevented if this occurs, but at least the processor is given an opportunity to correct a situation which allowed the illness to develop; a case in point was the large milkborne outbreak occurring in the Midwest in 1985. At some times, critical control points are not obvious and considerable research is required to establish the appropriate points where corrective measures may be taken. The type of research needed in such situations may consist of statistically designed and valid sampling programs, perhaps even repeat samplings, which will detect those points where contamination or failure

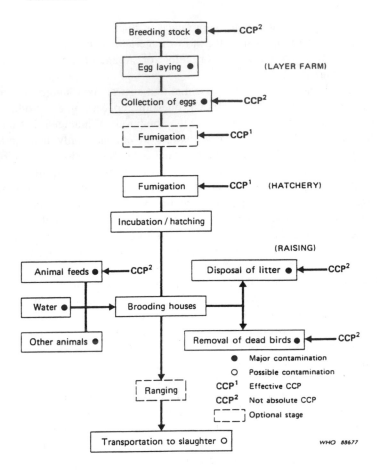

FIGURE 6.1   Sources of contamination and critical control points in poultry husbandry. (From Salmonellosis Control: The Role of Animal and Product Hygiene, Tech. Rep. Ser. No. 774, World Health Organization, Geneva, 1988. With permission.)

occurs and where corrective measures may be taken. The points detected in these studies would not be concerned with the appearance or quality of the product, but entirely with the hazards involved in the production or permitting of disease outbreaks. These points may well be crucial to the maintenance of a satisfactory shelf life for the product. When such studies are carried out within a plant, the final product should be reviewed by qualified regulatory personnel — in the case of meat processing, personnel of the Food Safety and Inspection Service of the USDA.

Once critical control points are recognized and established by hazard analysis, they must be monitored by qualified personnel on a regular or periodic basis as required by the hazard identified. Although the monitoring program will

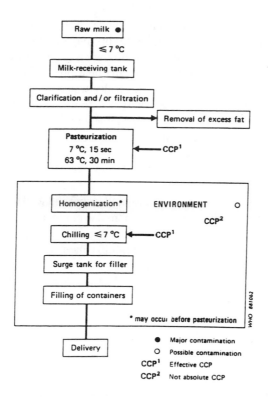

FIGURE 6.2 Sources of contamination and critical control points during the processing of milk. (From Salmonellosis Control: The Role of Animal and Product Hygiene, Tech. Rep. Ser. No. 774, World Health Organization, Geneva, 1988. With permission.)

most likely be done by plant personnel, it must be supervised and verified by FSIS personnel if it is to be effective. Failure of this verification results from a totally voluntary program which is not universally successful as discussed earlier. The kind of monitoring program used will depend upon the product, the process, and the hazard analysis which determines the critical control points. In all cases, however, the monitoring program should begin with the raw materials used in the product. Depending upon the operation of the plant, and the consistency with which raw materials are obtained from the same suppliers, it may be possible that the supplier carries out the monitoring program on the raw materials and there is no need for the user to repeat the process in all cases. In any event, the user should inspect, and depending on the nature of the product, monitor the raw materials coming into the plant to assure that they remain of high quality in all cases. Certainly tests should be carried out on any shipment from a new supplier for the first time. The monitor program may well consist of only physical and/or chemical tests, but in some cases may require the use of microbiological testing to assure safety. For example, in a process where heat-stable toxins may be a hazard, it is necessary that the raw materials be tested microbiologically to assure that toxigenic microorganisms are not present prior to the heat processing points. Finished products most often are not microbiologically tested because examination to determine pH, water activity, preservative activity, and salt level will give more information about the quality and the stability of the product than will general microbiological testing. In some instances, such as a roast beef product which is not to be further heated, it may better assure safety if the product is tested specifically for *Salmonella* contamination. Such a HACCP program efficiently carried out in a plant should give a margin of food safety that is highly desirable if it is carried out by knowledgeable and well trained-personnel. If such training has not been given, then the entire program may well be wasted.

A HACCP program for the meat and poultry industries may be generalized, although many of the critical control points (Figure 6.3) may well be slightly different from those found in other food industries. A major difficulty receiving publicity in recent months has been the use of antibiotics as growth-stimulating substances in animal feeds of all types. Consumer groups have complained about excessive concentrations of antibiotics in most animal products from milk and dairy products to poultry and red meats. This critical control point is found in the animal production process, and many are convinced that it presents a very real health hazard for the consuming public. That hazard may be in the form of production of resistant pathogens in the animals which are then transmitted to man, or in the form of consumption of subtherapeutic concentrations of the drugs which interfere with the treatment of infectious diseases in the human. The timing of application and the kind of antibiotic used on a farm makes a difference in the hazard posed to the ultimate consumer in the form of resistant bacterial pathogens, as well as in the form of antibiotic residues to the consumer. The use of antibiotics in farm animals to treat infections or to promote growth will not remove the danger of carriers of salmonella, and probably not of many other human pathogens. If the carriers are present in the farm flocks or herds when shipped to market, the pathogens will be carried into the slaughterhouses and result in contamination of that processing environment, transfer to the carcasses, and eventually to the meat products. In poultry production, the use of salmonella-free feed can have a major effect on the control of spread of this pathogen in the flocks and on to the carcasses and the consumers. In this or other production operations, once the pathogen is established in one animal, then the feed used, the water supplied, and the environment in general is likely to be contaminated soon.[6] A very recent finding of research in poultry production is that the addition of lactose, milk sugar, to the drinking water blocks the establishment of *S. typhimurium* in the intestinal tract of the chicken. If successful, the current thought is to attempt to use this sugar as an additive to the feed of the chicks, which would provide a fast, inexpensive control for this critical control point

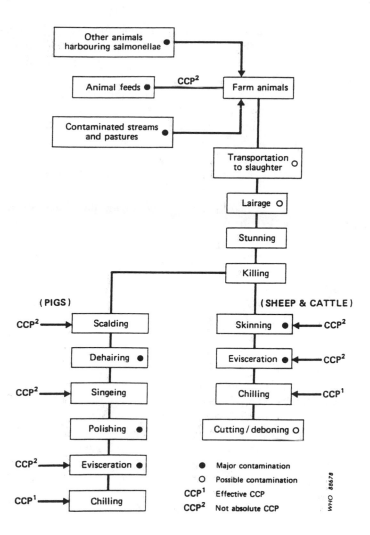

FIGURE 6.3 Sources of contamination and critical control points before and during the slaughter of pigs, sheep, and cattle. (From Salmonellosis Control: The Role of Animal and Product Hygiene, Tech. Rep. Ser. No. 774, World Health Organization, Geneva, 1988. With permission.)

in the production of poultry.[8] What is now needed is similar additives, with rapid results, for control of other dangerous salmonella serovars and pathogenic bacterial species.

The slaughtering-dressing operations in animal processing constitute critical control points which determine the final bacterial counts on the meat carcasses and products. Important within these operations at all times are the time-temperature relationships discussed generally above. During these processes, the microbial populations present are critically important, and control measures must be instituted to avoid spread of additional contamination during plucking poultry and skinning or dehairing swine. The general health of the animals and the sanitary conditions for transportation can largely be detected visually for many conditions if the animals were subjected to certain diseases. For microbial population identifications, however, the products must be subjected to microbiological cultures. The temperature of the scald water for both poultry and swine are critical controls to be used at this stage of the process, and if the time-temperature relationships do not hold, then the carcasses will proceed through the process with added or increased microbial counts, providing additional sources for cross-contamination points farther along the production program. Poultry carcasses after being plucked and swine carcasses after dehairing should be washed carefully, and again the time-temperature relationships are critical to avoid transfer and cross-contamination with salmonella, as well as other pathogens.

In addition to the drug residues in meat products, pesticide residues may present a health hazard to some individuals. Many pesticides are directly toxic to the human who consumes excessive amounts, and it has been shown in some cases that potential pathogens, including the salmonella, are capable of metabolizing some pesticides.[9-12]

The surfaces of all meat cuts and the surfaces and cavities of poultry carcasses are usually contaminated by a wide variety of microorganisms, including low numbers of some pathogens such as salmonella. This contamination comes from a number of sources, including the workers hands, cutting boards, table tops, saws, knives, and any equipment which has been wiped with cleaning cloths. Room temperatures where the meats are being handled and processed are critical in allowing these contaminants to multiply to a level which may prove dangerous, and in this manner become critical control points in the process. When raw meats are being ground, the contamination already present is included and receives additions from the grinder. If trimmings are not handled properly and removed frequently, these can become a source of added microbial contamination because the room temperature where the work is done may have allowed considerable growth of contaminants on the surfaces. The trimmings should be stored at a temperature no higher than 0°C until it can be permanently removed from the work site. As soon as practical, the ground or cut product should also be stored at near freezing temperatures to inhibit the growth of microorganism contamination. At this point in the process, the critical control points are perhaps more important because there will be less opportunity for correction of conditions following this for many of the products. When these products are further processed by vacuum packing or curing, additional critical control points are present and must be carefully monitored. The integrity of the packaging in vacuum packing is most critical to keep the carbon dioxide or nitrogen atmospheres adequate to inhibit the growth of aerobic contaminants which may be present. This control point must be regularly monitored to avoid spoilage of products, but does not remove the critical nature of the temperature control point in the storage of the packed product.

The actual process used in curing or fermenting meat products must be carefully analyzed for hazards and the critical control points located. Since different processes are sometimes used in different plants, these points must be specifically located in each instance, although some generalization can be made for points in such processes. In all processes and all plants, the beginning quality of the meat is the first and most important control point. If low quality product is used initially, the amount and type of contamination already present will most probably result in an unacceptable end product. In curing processes, the concentrations of the curing chemicals (brine, nitrite, nitrate) comprise a critical control point, as does the heat treatment to fix product color, but which does not kill microbial cells present. Storage time-temperature relationships add another critical control point which must be monitored and controlled to allow application of corrective measures when needed. In fermented products, the correct and timely addition of the fermenting culture are perhaps the most critical control points. The environmental conditions must be carefully controlled to assure the rapid growth of the fermenting lactic acid bacteria cultures. When the products consist of sausages stuffed into skins, the temperature and pH of the product during and immediately after stuffing become critical control points. At this point, monitoring must be accurate and constant, and reducing the temperature before fermentation may help to reduce the risk involved. If the product is smoked, this reduces the hazard of microbial growth by drying and thereby reducing the water activity of the product, but this process alone does not kill the vegetative cells of pathogens. For that purpose, the application of sufficient heat is required in addition to the smoke process. While water activity will prevent the growth of organisms it will not rapidly kill these cells, and although not growing, these cells can be revived and multiply when environmental conditions are altered in the storage or handling of the product.

Dried meat or poultry products present different critical control points for the prevention of hazards. In these processes, again, the initial quality of the meat and monitoring for time-temperature relationships are most critical, and hazards are most dangerous during the drying process while moisture content is being lowered. Once moisture content is lowered sufficiently, the storage stability and shelf-life have been extended to protect the quality of the product protected from reabsorption of moisture. The critical control point in that case is the integrity of the packaging. Temperatures during drying are not high enough to kill pathogens present, and again control of rehydration in the reconstituted product must be maintained by carefully maintaining the correct time-temperature relationships.

For cooked or pasteurized meat and poultry products, the cooking operations should be maintained completely separated from raw processes. In general the critical control points in these processes are not greatly different from those mentioned before. These include first and foremost, the quality of the raw ingredients and the time-temperature of the cook. Additionally, however, the time-temperature or rate of cooling after the cook must be added as critical control points, as must be the general sanitary procedures of handling and managing any equipment used. Sampling and microbiological cultures after heat processing is an excellent way to monitor these critical control points. These results for aerobic pathogens will not monitor for anaerobic, spore-forming bacteria, and the heat used is not sufficient to kill spores of such pathogens as *Clostridium perfringens*. If the cooked product is to be further processed, as in repackaging, boning, or slicing, each step becomes a critical control point which must be monitored. Cooked products must be cooled rapidly enough to prevent germination of spores and growth of vegetative cells, and for this purpose the chilling water should be properly chlorinated to prevent cross-contamination between products and containers. Canned meat products must have constant monitoring of the time-temperature relationships and pH to avoid dangerous hazards. If the products have a pH of 4.6 or below, they are considered high acid, and the GMP for these products is mandatory. If those procedures are followed, then critical control points and hazard analyses are specified, and the plant personnel do not have these decisions to make. If the products are low acid products (pH 4.6 and above), then the heat applied must be sufficient to kill *Clostridium botulinum* spores. Once canned, samples of the product are often incubated at elevated temperatures for a period of days to assure that shelf-life is adequate, and that the product is safe. Meat and poultry products processed by other means (irradiation, commercial sterilization, etc.) have some differences in critical control points and must be carefully analyzed.

Former FSIS Administrator, Donald L. Houston[13] stated that the concept of HACCP is a part of the strategy of FSIS, particularly in reference to "the Department's procedures for approving quality control plans in processing plants...". If this is to be the case, persons in charge of quality control operations and HACCP programs must have the proper educational training in HACCP principles, food science, including microbiology, and the details of food technology involving the processing operations. Inspectors need at least the basics of that training and the skills to monitor and analyze the hazards, as well as the critical control points.

Bryan[14] in 1978 analyzed 1152 outbreaks of foodborne illnesses and determined that by far the majority of them resulted from some breakdown in food service establishments or the home.[15] Of those which did occur as a result of breakdowns in food processing establishments, the two causes of the outbreaks which were most common were contaminated raw foods or ingredients and inadequate thermal processing. These two defects accounted for 50% of the outbreaks resulting from food processing establishments. Another 32% of the outbreaks resulted because of improper cooling and faulty fermentations in the processes. Infected persons and obtaining raw materials from unsafe sources accounted for the remainder of the outbreaks. Had a HACCP system been in operation at these food processing plants, it is likely that these defects would have been detected, and the outbreak would thereby have been avoided without intervention by an inspector on a daily basis.

The International Commission on Microbiological Specifications for Foods,[15] through the World Health Organization,[16] has recommended a HACCP program for the effective prevention and control of salmonellosis. Such a program is represented by the Figures 6.1 to 6.3. In these figures, CCP[1] represents those control points where most effective measures may be applied for the prevention of contamination, and CCP[2] represents those points where, although control measures are applied, they may not always be 100% effective. In these latter cases, it is most important to continue vigilance and to assure that the points are monitored at all times. With proper monitoring and corrective actions, these points are greatly reduced hazards.

Workers in food processing or production plants are often the immediate sources of pathogenic organisms, including *Salmonella*. These workers, like others of us, are often infected with these organisms, and when this is the case, the organisms are at some times shed in the stool. In the case of food service establishments, many local and state governments have at some time, and some still do, required physical examination, including stool cultures to assure that the workers are not shedding the organisms. If these pathogens reach the hands of the workers, then they are potentially dangerous to the products, because of the likelihood of transfer from hands to foods. Almost all ordinances prohibit ill persons from handling food in any establishment, but the problem comes in the determination of illness. With some frequency, there is an outcry through the news media for inclusion in food ordnances of a requirement for health examinations for workers in food industries, and particularly in those where food is served to the public. Where the requirement does exist, it generally states that the examination must be done at annual intervals or sometime as often as six months. These requirements frequently include testing for syphilis and for tuberculosis, neither of which is likely to be transmitted by infected persons handling foods. Such testing can be valuable as screening programs, but will have no effect on the safety of foods.

On the other hand, the inclusion of stool culture examinations to detect the presence of *Salmonella* in the feces of workers may be significant *at the time it is done*, but it will not be significant at a later time, even tomorrow. In fact the failure to find the pathogen in a single culture is not significant because even when infected an individual may shed the organism only when that person experiences diarrhea, and the shed of the pathogen may be intermittent. Since many foods of animal origin may be externally contaminated by *Salmonella*, the worker may leave the site of the health examination, return to work, and within the hour pick up the pathogen, thereby initiating an infection. In most instances, the carrier condition exists only following infection with *S. typhi* or one of the paratyphoid serovars, but it can occur with other *Salmonella* types as well. In some cases a carrier may be detected by a single stool culture, but in many instances several cultures, repeated at intervals of 2 to 3 weeks may be necessary to detect the pathogen. The detection of a carrier of any pathogen, particularly *Salmonella*, is important to the food industry, because it may prevent future contamination of products which would otherwise be unsuspected, whereas the detection of an infection may prevent trouble today, but the danger will be gone within a few days as the infection is ended.

When cultures are negative in these health examinations for any or all pathogens, the workers in a food industry may well be misled as to the possibility that they can transmit pathogens to food, because they can so easily be infected after the examination, and with mild or no symptoms may well never know that they are infected. When such a program exists and is administered through a health agency, and no positive cultures are detected, the agency may develop a false sense of security in that they may feel that the public is being protected, when in fact there is no positive result of the program. An intensive program of teaching and reminding of the necessity and the benefits of frequent and proper handwashing (as discussed in Chapter 3) may well pay better dividends than a health examination program involving tests at intervals of several months. The program is not likely to involve testing more frequently than this because of the high cost involved in these examinations.

The National Advisory Committee on Microbiological Criteria for Foods has recently given final approval to its first major project — Hazard Analysis and Critical Control Point System.[6] This document has been forwarded to the major agencies dealing with food safety — the Food Safety and Inspection Service of the USDA, the FDA, the National Marine Fisheries Service, and the Army-Natick Research Development and Engineering Center. In this work, it is stated that HACCP is an effective and rational approach to the assurance of food safety. Among the principles stated in this work is that the critical limits that must be met at each critical control point must be established or identified. It is also stated that corrective action must be established and taken whenever there is a deviation from the ideal at a critical control point. It is pointed out that because of the differences in food processing and the variations in critical control points in those processes, the specific corrective actions must be developed for each point in the HACCP plan for each food plant operation. Finally, the HACCP plan must establish verification that the plan is working correctly to accomplish food safety.[17]

REFERENCES

1. **Bryan, F. L.,** Impact of foodborne disease and methods of evaluating control programs, *J. Environ. Health*, 40, 315–323, 1978.
2. **Kauffman, F. L.,** Interpretation and application of FDA laws and regulations, in *Control of Critical Points in Food Processing: A Systems Approach*, Doyle, E. S. and Mittler, A., Eds., The Bosley Corporation, Washington, D.C., 1977.
3. An interview with Dr. Lester Crawford, *Food News for Consumers*, Food Saftey Inspection Service/U.S. Department of Agriculture, 6(4), 6, 1990.
4. **Silliker, J. H.,** Principles and applications of the HACCP approach for the food processing industry, in *Food Protection Technology*, Felix, C. W., Ed., Lewis Publishers, Chelsea, MI, 1987.
5. **Food Marketing Institute,** *HACCP Manual*, Food Marketing Institute, Washington, D.C., 1989.
6. **Committee on Food Protection,** *An Evaluation of the Role of Microbiological Criteria for Foods and Food Ingredients*, National Research Council, National Academy Press, Washington, D.C., 1985.
7. **Committee on the Scientific Basis of the Nation's Meat and Poultry Inspection Program,** *Meat and Poultry Inspection*, National Academy Press, Washington, D.C., 1985.
8. **Schor, D.,** The FSIS science challenge — keeping pace with new products, new microbes, *Food News for Consumers*, 6(4), 8, 1990.
9. **Murray, H. E. and Guthrie, R. K.,** Effects of carbaryl, diazinon and malathion on native aquatic populations of microorganisms, *Bull. Environ. Contam. Toxicol.*, 25, 535, 1980.
10. **Murray, H. E. and Guthrie, R. K.,** Metabolic responses of aquatic bacterial populations to selected insecticides, *Water Resour. Bull.*, 16, 749, 1980.
11. **Anugwelem, U. A., Guthrie, R. K., and Davis, E. M.,** Response of bacteria to the presence of carbaryl in water. I. Response of bacterial indicators vs. response of pathogenic bacterial species, *Water Resour. Bull.*, 17, 1000, 1981.
12. **Anugwelem, U. A. and Guthrie, R. K.,** Measurement of effects of carbaryl in surface water on fecal pollution indicator bacteria by use of plate counts and respirometry, *Water Res.*, 16, 219, 1982.
13. **Houston, D. L.,** Quoted by Dr. L. Crawford, *Food News for Consumers*, 6(4), 8, 1990.
14. **Bryan, F. L.,** Control of foodborne diseases, in *Safety of Foods*, Graham, H. D., Ed., AVI Publishing, Westport, CT, 1980.
15. **Graham, H. D.,** *Safety of Foods*, 2nd ed., AVI Publishing, Westport, CT, 1980, 265.
16. **WHO,** Salmonella Control: The Role of Animal and Product Hygiene, Tech. Rep. Ser. No. 774, World Health Organization, Geneva, 1988.
17. American Association of Food Hygiene Veterinarians, *News-O-Gram*, 14(1), 1990.

Chapter 7

SALMONELLA ENTERITIDIS IN EGGS

In 1942, there were approximately 5500 isolates of *Salmonella typhi* in the U.S. That number of isolates has been reduced to below 500 currently. On the other hand, other *Salmonella* serotypes have been isolated at an increasing rate, from below 500 in 1942 to between 10,000 and 60,000 isolates in 1985.[1] These changes have resulted from the changes in food habits in this country, as well as in changes in production technology for all foods. The serotypes which have been isolated from certain foods have changed in frequency of isolation as well. For example, since 1986, the isolations of *S. enteritidis* has surpassed the isolations of *S. typhimurium* (a most frequent isolate in many states). *S. heidelberg* isolations were parallel to those of *S. enteritidis* up to 1986, but the isolations of SE have been more prevalent since that time.

S. pullorum, a pathogen for poultry was most detrimental to the poultry industry because it caused many deaths among poultry flocks. The industry has essentially brought this organism under control in recent years.[1] That control was aimed at the transovarian transmission of the bacterium, and because of that success, it appears that a control program aimed at the transovarian transmission of SE in recent outbreaks in the U.S. would have equal chance of success.

As recently as the 1960s one of the most common means of transmission of *Salmonella* to susceptible individuals, when a food vehicle could be identified, was by way of contaminated chicken eggs. After considerable study, it was determined that the eggs were being contaminated by salmonella which were present in the chicken feces found on the outside of the egg. This fact emerged when it was found that the cases of salmonellosis produced were caused by different serovars or species of the pathogen, and the control of such infection was a relatively simple matter of controlling the amount of contamination on the shell of the egg, and in educating the cooks to break the eggs in a manner that would not contaminate the egg from the shell.[2]

Between 1975 and 1987, the Centers for Disease Control (CDC) in Atlanta determined that there had been a severalfold increase in the numbers of infections caused by *S. enteritidis* in several states in the Northeastern section of the U.S. In over 70% of the cases of infection noted, the source of the infection was traced to contaminated eggs, much as in the case of the contaminated ice cream product in the New York outbreaks. Actually, in 35 of 65 outbreaks of infection caused by these organisms, it was determined that uncracked eggs or foods that contained eggs which were incompletely cooked were involved in the transmission of the organisms.[3] In recent cases, there is no evidence of the presence of checks, or hairline cracks in the eggs, and contamination from the shell surface which had been in contact with fecal matter in the intestinal tract of the hen was ruled out. When CDC studied the data of the occurrence of *S. enteritidis* infection outbreaks in the Northeast between 1973 and 1987, it was concluded that most of the outbreaks occurred during the summer months when warm temperatures may have provided opportunities for the bacterium to multiply during production, transport, or storage.[4]

After the mid 1970s, however, it was noted that the incidence of salmonellosis caused by *S. enteritidis* had increased more than sixfold in the northeastern section of the U.S. Generally, only the outbreaks of multiple cases were investigated rather than the single cases which constitute the majority of reported salmonella infections in this country. It was determined that the majority of the outbreaks investigated were associated with consumption of eggs. In the period from 1985 to mid-1987 there were 65 outbreaks of *S. enteritidis* in that part of the U.S., involving 2119 cases and 11 deaths.[5] The food vehicle was identified in 35 of the outbreaks as Grade A eggs or foods containing these eggs. After much study these outbreaks of salmonellosis from carefully inspected and graded eggs were determined to be due to passage of the pathogen from the hen to the egg by transovarian transmission. In the first 10 months of 1989, 49 such outbreaks were reported, involving 1628 cases and 13 deaths. It is worthy of note that 12 of the 13 reported deaths occurred in nursing homes.

S. enteritidis has a wide host range, with some evidence that some strains are developing a predilection for poultry. This serovar, however, has been isolated from reptilian, avian, and mammalian species, and rodent reservoirs are often reported in zoos. Although the majority of cases and outbreaks have occurred in the Northeast or mid-Atlantic areas, there have been isolates and outbreaks thought to be associated with Grade A eggs in other localities, including Tennessee, West Virginia, Utah, and Colorado. There is speculation as to whether these increases are manifestations of the spread of SE in poultry flocks, particularly breeding flocks, which then permit rodents and other animals to spread the organisms in this country.

Surface contamination may reach the internal portion of the egg, and thus be transmitted to the consumer if the egg is not thoroughly cooked. The two modes of transmission of salmonella by eggs are noted in Figure 1.4. Consumption of raw eggs or eggs not completely cooked, i.e., "sunny side up" or eggs which have been inadequately boiled, may serve to provide the consumer with a sufficient dose of the organism to initiate an infection. In addition, eggs used raw in such foods as salad dressings, hollandaise sauce, eggnog, or ice cream as in the incidents described which occurred in New York in the 1960s, will serve to transmit the organism to the consumer. In the case of transovarian transmission, the shells of the eggs need not be cracked for the organisms to be present in the food and for the infection to be transmitted by foods.

Transovarian transmission makes the control of these infectious diseases much more difficult, and should such transmission occur with other pathogenic organisms, intensive efforts toward control of transmission will be required.

It is obvious that any contamination which occurs as a result of the organisms being carried by eggs could result from a case of transovarian transmission or from a case of shell contamination by passage through the intestinal tract. It is thus obvious that any of the outbreaks which have been investigated and found to result from the use of contaminated eggs may well involve either one or both of these situations; this includes the outbreaks reported from consumption of imitation ice cream in the New York incidents, and also includes the recent outbreaks involving scrambled eggs, omelets, hollandaise sauce, etc.

The increase in incidence of *S. enteritidis* infections, associated with the consumption of raw or undercooked eggs, has not been observed in the U.S. alone. Instead large increases in incidence have been reported from Yugoslavia, Finland, Sweden, Norway, and the U.K. In Britain alone (England and Wales) the number of cases in 1988 (15,427 isolations from human sources: 12,522 isolations of *S. enteritidis*, Phage type 4) was more than double the number of cases in 1987. *S. enteritidis*, Phage type 4 has not yet been detected in the U.S., but the mechanism of spread in this phage type, just as occurs in the types spread in the U.S. is also thought to be transovarian transmission from the hen to the egg (vertical transmission), and in some cases at least, on to the progeny. This phage type has caused measurable losses in chicken flocks in the Iberian Peninsula as well as in Great Britain, and extensive human illness is associated with this specific organism there as well. The U.S. Department of Agriculture has enacted restrictions to prevent either poultry or poultry products infected with this phage type of *S. enteritidis* from entering this country.

In the U.S., Phage type 8 has been recognized as producing grossly recognizable lesions in chickens, and this type has been observed to be transmitted by transovarian transmission at a high level in experimental work in this country. Other isolates have not been observed to have the same high level of vertical transmission or to produce such extensive disease in poultry flocks. The geographic spread of SE emphasizes the need for control programs to limit the spread of the bacteria in poultry flocks. In one situation, three egg producers in a region of considerable spread were found to be the likely sources of at least seven different outbreaks. The outbreaks, for some reason not yet determined, generally appear outside of the winter months. There are also believed to exist some strain differences in human pathogenicity among the isolates in this country, in addition to the observed differences in strain pathogenicity for the human and poultry, and in the apparent prevalence or number within eggs which have been vertically contaminated.

Vertical transmission (transovarian) from hens to progeny or table eggs has been well documented for SE. Such transmission rapidly increases the incidence of disease in poultry flocks and permits cross-contamination due to contamination of the egg passage through the chicken. It is thought that if the egg is contaminated during passage from the cloaca or oviduct, then the organism may penetrate the shell and still result in transovarian transmission in later generations produced by breeder eggs. Horizontal transmission also occurs to pass SE from contaminated poultry flocks to other poultry. The premises where contaminated poultry has been raised can expose susceptible flocks to SE from the poultry house, service equipment, or from personnel who are carrying the organism. A very common source for horizontal transmission of the organism is feed, particularly feed which contains protein by-products which may have been prepared from animals carrying the organism although the FDA did a survey of rendered poultry products in 1989 which showed no *S. enteritidis* in the products, though other salmonella serovars were found. Even when prepared from uncontaminated animals, the process of milling may bring the feed ingredients into contact with contaminated equipment or containers, which may help to explain the presence of some of the other types. The FDA found some contamination by other serovars in all raw material and in the finished products of all manufacturers in that survey.[6] Contamination of feed may occur in milling, storage, transportation, or even after the feed reaches the farm. Animal contamination in horizontal transmission may occur from people working or visiting the farm, from pets, wildlife, and particularly from rodents which may be able to reach the facility. When poultry dies, the carcass should be removed immediately to prevent the multiplication of any pathogen which may have caused the death or of any incidental pathogen which may have been present in or on the chicken. If environmental cleaning and disinfection is regularly and efficiently practiced, and if the operation strictly follows all-in, all-out operation practices, horizontal transmission can be reduced in daily operations.

At present, there does not appear to have been any reduction in the volume of the egg or poultry market because of the incidence of *S. enteritidis* outbreaks. On the other hand, the severe outbreak in 1988 in Britain has been estimated to have resulted in a permanent 20% loss in volume of the egg market.[7] Because of the magnitude of the problem which is now recognized with infection of people by *S. enteritidis* and because of the intensity of studies involving these outbreaks in recent years, this particular infection deserves special emphasis. If for no other reason, the attention attracted has resulted in more rapid, and more successful, efforts at control of spread of the infection than has occurred in the case of other types of salmonelloses. As was seen in Britain, outbreaks of such magnitudes can result in dramatically decreased demands for poultry and poultry products, and control becomes essential for industry stability.

Salmonella enteritidis serovar enteritidis, Group D shows signs of becoming strongly adapted to the chicken and the human. The organism produces an infection in chickens, but the symptoms are not highly specific. For this reason, the U.S. Department of Agriculture (USDA) has instituted testing programs (initially the programs were voluntary) to detect flocks infected and to begin the control of this organism and the resulting infection. In 1989 after a period of review and in response to increased outbreaks of SE infection, the USDA Animal and Plant Health Inspection Service presented a plan which recommended the establishment of the Salmonella Enteritidis Task Force[7] to work toward programs to control the spread of this infection in poultry flocks, and thereby to control the spread of the infection to the human by way of eggs and other contaminated poultry products. The Task Force establishment made it possible to meet the following basic needs: (1) to develop guidelines for the implementation of the regulation, (2) to assist operations in outbreaks and in testing of study and test flocks, and (3) to serve as a national information center for *S. enteritidis*. The successful functioning of the Task Force will aid in achieving the confidence of the public about the nutrition and safety of eggs, poultry, and poultry products. To this end, detection of breeding flocks or egg production flocks which are infected becomes critical, as are the operational biosecurity measures utilized by the poultry industry. Once detected, infected flocks are allowed only for the production of eggs for the pasteurization market. When an infected flock is discovered and removed from a production facility, it becomes necessary to carry out stringent disinfectant procedures because it has been amply demonstrated that the bacterium can persist in a viable form for up to 7 months in litter or soil, 28 months in manure, and up to 5 years in hatchery fluff.

Rather than these stringent requirements, it is much more practical to prevent salmonella, including SE infection in a poultry flock. The Salmonella Enteritidis Task Force[7] has urged producers to follow the biosecurity practices listed as follows:

- Clean and disinfect premises thoroughly between flocks.
- Clean and disinfect all vehicles and equipment entering and exiting the farm premises.
- Keep out unnecessary visitors and avoid borrowing equipment.
- Provide sanitized coveralls and boots for personnel entering poultry premises.
- Purchase replacement birds from breeders certified by the "U.S. Sanitation Monitored" program.
- Practice all-in, all-out poultry management. Birds left behind may carry disease to incoming flocks.
- Avoid contact with game birds, migratory water fowl, and rodents — suspected carriers of poultry diseases.
- Provide only quality, salmonella-free feed.

The causative organism identified in these outbreaks is *S. enteritidis* serotype enteritidis (SE), Group D in the Kauffman-White schema. This bacterium is a pathogen for the human, poultry, and other animals. In this respect, it is unlike *S. typhi* which has adapted to become a human parasite, which made transmission control of *S. typhi* somewhat easier. Other than the mortality rates, the symptoms of *S. enteritidis* infection in the human are not unlike other salmonelloses. The infection is usually followed by diarrhea, headache, abdominal pain or cramps, nausea, fever, and vomiting, or some combination of these symptoms. In this type of salmonellosis, the persons at greatest risk are the elderly, the very young (under 3 months of age), pregnant females, and persons suffering from other disease or who are immunocompromised in some way. While the infection often results in death in the elderly, it can be fatal in otherwise healthy adults when it is ingested in sufficiently large numbers. Although the overall fatality rate falls below 1%, the rate in the elderly may be considerably higher, in some cases reported as high as 8%.[5] Septicemia has been reported in these infections, particularly in the elderly. Apparently, the fatality rate in otherwise healthy adults is somewhat dependent on the dose of organisms ingested to begin the infection. Because the infecting dose may be larger with the incubation of the organisms in the eggs at room temperature, the incubation period may be slightly shorter than is seen in other salmonelloses, ranging from approximately 6 to 72 h, although the latter time is longer than the usual incubation period. This bacterium, more commonly than other serovars of *Salmonella*, invades organs outside the intestinal tract and because of this characteristic may result in more serious complications. Complications which have been reported, both in experimentally infected rats and in the human, include chronic reactive arthritis in 2% of the patients, but more often consist of focal infections in various organs.

In localities where egg-associated salmonellosis has been identified, campaigns should be effected to educate the public that foods containing raw or undercooked eggs should not be eaten. In preparation of foods containing eggs, a single SE-contaminated egg (among many used) can cause outbreaks of severe illness. Foods containing liquid or runny egg materials can contain living *Salmonella* and can cause infection in the consumer. Such foods prepared and allowed to stand at room temperature for more than 2 to 4 h are likely to cause infection in this manner. Thorough cooking of the foods will kill any *Salmonella*, including SE which are present. Preparation of eggs for human consumption in any setting, and particularly for consumption by those most susceptible to infection, should strictly follow guidelines for cooking which have been proven safe. Pasteurization of contaminated eggs will kill any *Salmonella* present, and when foods requiring raw or undercooked eggs are required this product should be used. Although pasteurized eggs are not always easily available to individuals, bulk pasteurized eggs are available commercially and food service establishments (including health facilities) are advised to use the pasteurized rather than the raw product. Service of raw eggs and foods containing raw eggs such as Caesar salad, Hollandaise sauce, ice cream, eggnog, and mayonnaise should be avoided, as should lightly cooked foods such as soft custards and French toast unless pasteurized eggs are used in preparation. Eggs should be cooked until the yolk and white are both firm, not runny. Generally a temperature of 250°F is required for safety in cooking eggs. Further information on cooking and handling eggs safely for individuals or for food service establishments may be obtained from the Department of Agriculture Meat and Poultry Hotline, and in most states from county extension home economists.[8]

Egg handling practices should include sanitary precautions just as should handling of raw poultry.[9] Before and after handling eggs, hands should be washed with warm to hot soapy water, and any equipment used such as containers, blenders, or beaters should be washed and sanitized between uses. When eggs are purchased for home use and stored in the refrigerator, they should be left in the original carton and stored in a cooler part of the refrigerator since the door with the egg section may not be at correct cooling temperatures in most home refrigerators and will permit slow multiplication of bacteria, including the salmonella. Perhaps one of the areas of greatest danger in commercial service of eggs, is the use of a steam table of improper temperature for serving scrambled eggs. If cooked in large quantities, the scrambled eggs may still contain viable bacterial cells in the softer or runnier portions, and addition of these to inadequate temperature steam tables allows incubation and growth of the microorganisms.

The most recent *S. enteritidis* concern is not the first time that the egg industry has been threatened by outbreaks of salmonellosis. In the 1960s, outbreaks of the disease associated with eggs occurred and were influential in changes in the operation of the egg industry and establishment of standards of cleaning and disinfecting shells of whole eggs. Rather than concern about transovarian transmission, those contaminations were generally the result of fecal contamination on the shell, and of contamination in shell cracks with seepage of the egg from the crack. At that time, the causative agent of greatest concern was not *S. enteritidis*, but rather one of a number of other organisms, including the primarily chicken-adapted strains of *S. pullorum* and *S. gallinarum* was likely to be involved. The transmission mechanism involved included contamination of shells with the pathogens carried in fecal material which could then migrate internally or contaminate the egg in the process of breakage. By disinfecting and washing, great success has been observed in restricting the transmission of the two types of salmonella listed above, as well as other serovars of salmonella. The two serovars primarily adapted to poultry are now said to have been eliminated in 36 states and have been greatly reduced in other states. These efforts at reducing infection have been made through the National Poultry Improvement Plan, which continues to test flocks in the effort to eliminate disease from the most critical source — the poultry breeding stock[10] from which the organisms and disease spreads to the human by way of eggs or poultry meat.

The seriousness of the recent spread of SE in the U.S. is pointed out by Altekruse.[11] The difficulties in the past few years have resulted in new regulations and restrictions on movement of poultry and poultry products by the USDA. These regulations depend heavily upon the flock definition categories with respect to the presence of *S. enteritidis* in the birds or in eggs. The flock definitions are found in the February 16, 1990 *Federal Register*[12] summarized as follows.

A. STUDY FLOCK:

Study flock status is achieved in the following ways:

1. Any egg-production flock reported to be showing clinical signs of disease
2. Any report implicating any egg-production flock as a "probable source" of a human disease outbreak
3. Any report implicating a breeder flock as the "probable source" of a disease outbreak in an egg-production flock
4. Any progeny placed since the last negative environmental sampling from a breeded flock which now has a positive environmental sample

The result of Study Flock status — Within 15 days of this designation by a federal or state authority, environmental samples from manure or manure scraping machinery and egg transport equipment must be taken by the state or federal representative from every row in each house on the ranch and submitted to an authorized diagnostic laboratory for culture.

Sanctions — No sanctions will be imposed on study flocks with regard to prohibition of interstate product movement, provided the owner and manager comply with the regulations permitting sampling and testing.

B. TEST FLOCK

Test flock status is imposed on the basis of the following:

1. Any positive environment sample
2. Refusal by the flock owner or manager to permit sampling within 48 h of notification of study flock status
3. Any action by the flock owner or manager deemed to have resulted in a delay in completion of the sampling beyond 15 days from notification
4. The determination by a state or federal representative that the flock is the "probable source" of three or more outbreaks of human disease

The result of Test Flock status — Three hundred blood samples will be collected by the state or federal representative from each house on the ranch. The 300 samples will be taken randomly; in addition, any clinically ill birds will be sampled. Sampled birds will be individually banded for later identification. Additionally, 60 birds from each house will be taken for internal organ culture from among seropositive birds and birds chosen at random from rows where positive environmental samples were collected.

Negative organ culture does not rescind test flock status; the testing is repeated in 15 days. Positive organ culture (all organs except lungs and gastrointestinal system) results in a designation of Infected Flock status.

Sanctions — Identical interstate movement restrictions will be imposed on both Test Flocks and Infected Flocks.

C. INFECTED FLOCK

Infected Flock status is based on a single positive organ culture sample.

Sanctions — Articles subject to interstate movement restriction include live chickens (movement for slaughter within 24 h only — permit required), eggs (movement to breakers for pasteurization only — permit required), manure (movement permitted in covered containers only, for burial, or use on fields not used for grazing or poultry production, or composting — permit required), cages, coops, troughs, and other equipment (movement allowed only if constructed of plastic or metal, cleaned and disinfected in the presence of a federal or state representative — certificate required).

Rescinding Infected Flock status requires retesting and complete absence of *S. enteritidis* on culture of environmental samples (manure, egg belts, machinery, etc.) and organ culture.

The new status will not be in effect (nor sanctions lifted) until the flock owner is notified in writing of the negative culture results.

The rationale cited for the plan is outlined as follows:

1. *S. entertitidis* serotype enteritidis:
    a. Is associated with clinical disease in poultry
    b. Is known to occur in the U.S.
    c. Has been isolated from egg-type, chicken breeding flocks, and egg production flocks
    d. Can be spread horizontally by direct contact and contact with articles associated with infected poultry, such as feed pens, and litter
    e. Represents a public health concern that shows no sign of abatement
    f. Is involved in infecting a number of commercial egg-laying flocks
    g. Has contaminated a quantity of commercial table eggs
    h. Is causing a growing number of cases of human illness and death

2.   The domestic egg industry:
    a.   Is organized pyramidially with approximately 900 primary and multiplier breeder flocks
    b.   Consists of the breeder flocks and approximately 3500 commercial laying flocks (these criteria also apply to turkey and broiler flocks)
3.   The National Poultry Improvement Plan:
    a.   Was amended in 1989 to require SE testing in participating breeder flocks (Docket #89-049)
    b.   Is being amended to create a new "United States Sanitation Monitored" classification for flocks which have met the SE testing requirements (all interstate shipments must come from such flocks)
    c.   Is now being amended to place APHIS in the reporting chain for SE (and all other laboratory-diagnosed poultry diseases) by requiring authorized laboratories to meet all of the following requirements:
        i.   Technical personnel must receive training prescribed by NVSL
        ii.   Reagents, media, and antigen must be approved by NVSL
        iii.   NVSL-approved protocols must be followed
        iv.   Tests must meet proficiency test check requirements
        v.   All test results must be reported to State officials and APHIS[12]

These control programs, including the testing requirements, are depicted in Figures 7.1 and 7.2.

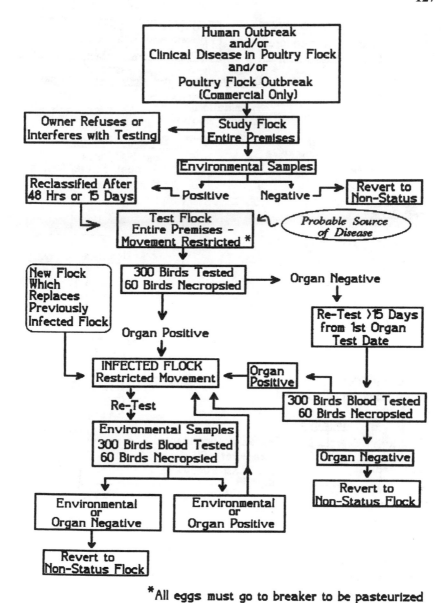

FIGURE 7.1 Sequence of testing, designations, and requirements for Task Force control programs. Flow diagram courtesy of Garth Morgan.

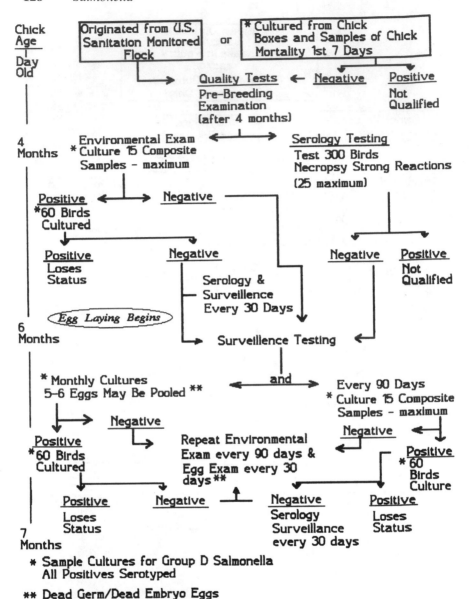

FIGURE 7.2 Sequence of testing and surveillance of the National Poultry Improvement Plan. Flow diagram courtesy of Garth Morgan.

REFERENCES

1. **Shipman, L.,** presented to *Salmonella enteritidis* Regional Work Conference Central Region, Kansas City, MO, 1990.
2. **Nightingale, S. L.,** Salmonella foodborne illness. An FDA perspective, *Am. Fam. Physician,* 39, 387, 1989.
3. **Waldholz, M.,** Rise in Salmonella Cases is Tied to Eggs Formerly Believed Free of the Bacteria, *The Wall Street Journal,* Friday, April 8, 1988, 13.
4. **Taylor, G. C., Meadows, M. A., Jargowsky, L. W., Pilot, K., Teter, M. J., Brook, J., Petrone, M. E., Spitalny, K. C., Spitz, S. B., Davin, R. J., Ellison, J., Kondracki, S. F., Guzewich, J. J., Fudala, J. K., Debbie, J. G., Morse, D. L.,** Update: *Salmonella enteritidis* infections and Grade A shell eggs — United States, 1988, *Morbid. Mort. Wkly. Rep.,* 37(32), 490, 1990.
5. **St. Louis, M. E., Morse, D. L., Potter, M. E., DeMelfi, T. M., Guzewick, J. J., Tauxe, R. V., Blake, P. A.,** The emergence of Grade A eggs as a major source of *Salmonella enteritidis* infections, *JAMA,* 259, 2103, 1990.
6. **American Association of Food Hygiene Veterinarians,** *News-O-Gram,* Vol. 14, January 1990.
7. **U.S. Department of Agriculture,** *Salmonella enteritidis* Task Force, U.S. Department of Agriculture, Washington, D.C., 1990.
8. **Steinert, L., Virgil, D., Bellemore, E., Williamson, B., Dinda, E., Harris, D., Schneider, D., Fanella, L., Bogacki, V., Liska, F., Birkhead, G. S., Guzewich, J. J., Fudala, J. K., Kondracki, S. F., Shayegani, M., Morse, D. L., Dennis, D. T., Healey, B., Tavris, D. R., Duffy, M., Drinnen, K., and Hutcheson, R. H.,** Update: *Salmonella enteritidis* infections and Grade A shell eggs — United States, 1989, *Morb. Mort. Wkly. Rep.,* 38, 877, 1990.
9. **Blumenthal, D.,** From the Chicken to the Egg, *FDA Consumer,* U.S. Department of Health and Human Services, Washington, D.C., April 1990, 7.
10. **U.S. Department of Agriculture,** Pathogen of People and Animals, Fact sheet, *Salmonella enteritidis,* U.S. Department of Agriculture, Washington, D.C., March 1990.
11. **Altekruse, S. F.,** The Transmission and Spread of *Salmonella enteritidis,* Central Regional Work Conference, U.S. Department of Agriculture, APHIS, Kansas City, MO, June 20, 1990.
12. *Federal Register,* Vol. 55, #33 (9CFR Parts 71 and 82 — Docket #88-161), February 16, 1990.

Chapter 8

MICROBIOLOGICAL METHODS FOR DETECTION OF *SALMONELLA* CONTAMINATION

Most methods used to culture the salmonella originated as methods for culturing clinical specimens, and these have since been modified to be useful for food and environmental samples. These methods and their modifications begin with the routines for pre-enrichment of samples, and follow through selective enrichments, selective plating, differential testing and plating, confirmatory biochemical determinations, and confirmatory serological testing.[1] A major difference in the methods required for these purposes has been the need for enrichment of samples of foods to allow the growth of salmonella from very small numbers which are frequently in the minority in the sample where they exist in foods, whereas, in clinical samples, these pathogens may predominate, or in some cases even be in almost pure culture. In certain foods, it is also necessary that some food components be neutralized before the bacteria will be able to grow. These food components may not be toxic enough to kill the bacteria, but their presence is sufficient to inhibit growth of the organisms, and thereby to prevent culturing.

Methods used to detect *Salmonella* contamination of raw or processed meats or poultry must ultimately be approved by the Food Safety and Inspection Service (FSIS)[2] of the U.S. Department of Agriculture (USDA). Methods used for other types of foods must be approved by the Food and Drug Administration (FDA), which primarily uses methods approved by and published by the Association of Official Analytical Chemists (AOAC). Many methods are first tested and approved by the AOAC. Frequently, methods and procedures stay in place for a good many years after approval because it is necessary that new methods be thoroughly tested and the results analyzed under differing conditions so that analyses can be relied upon when coming from the many environments of food processing and marketing companies, or from commercial laboratories engaged by these companies to do such testing. For the most part, methods utilized and approved by FSIS for meats and poultry will be satisfactory for other foods as well. Where there are exceptions to this rule, these cases will be noted in the methods section to follow. Frequently, differences which are recommended are due to the nature of the foods which indicates that a different enrichment medium or procedure would increase the chances for recovery of *Salmonella*.

There are several associations and/or agencies which publish microbiological methods for testing foods to detect the presence of *Salmonella*. These include the AOAC. For approval as an official method of the AOAC, the method must be reliable, practical, available to all analysts, and substantiated by repeated and repeatable results in use. These methods are included in the *Official*

Methods of Analysis.[3] The FDA includes their methods in the *Bacteriological Analytical Manual.*[4] The Centers for Disease Control published methods which they feel most useful in their investigations in 1968.[5] The USDA methods are included in their *Microbiology Laboratory Guidebook,*[2] to which methods are added as they are found useful and reliable. A number of international organizations concerned with microbiological criteria for foods are often involved in efforts to develop, improve, and standardize methods for the microbiological examination of foods. Among these organizations are the International Commission on Microbiological Specfications for Foods, International Association of Microbiological Societies; Codex Alimentarius Commission, Expert Committee on Food Hygiene; the Joint FAO/WHO Expert Committee on Food Hygiene; and Subcommittee 6 (Meat and Meat Products), International Organization for Standardization. In addition to these organization and agency efforts, there are frequent studies involving individuals or commercial companies designed to improve the efficiency, speed, and/or accuracy of existing or new technology methods.[1] When such methods become available, they must be repeatedly tested in many laboratories before they can be approved by any of the official or regulatory agencies. New methods are often evaluated by research laboratories in academic or regulatory settings, with the findings published in current literature, as was done by D'Aoust and Sewell[6,7] for the Bio-enzabead™ enzyme immunoassay and the immunodiffusion 1-2™ test System, and by other workers for these and other systems.[8-13]

In order to establish a Total Quality Control Program (TQC) in the meat or poultry industries, at least five control programs must be made mandatory: control of microorganisms, control of fat content and added water, control of net weights, control of processing temperatures, and control of chemical composition from the standpoint of detecting contaminant or added materials which illegally change the nature of the food. Such a monitoring system must be accomplished by verification samples obtained and tested by the FSIS, and examination of records of the processor on an annual or more frequent basis. The records and routine testing between inspection tests of the FSIS must remain the purview and the responsibility of the plant manager, and the production manager from the company producing the food should not be involved. Obviously, small plants are going to be unable to afford the expense and expertise of a satisfactory TQC, and allowances must be made by the FSIS to permit testing and compliance of these operators by some other means. The TQC system establishes a relationship with industry operators which is essentially positive, rather than the more traditional faultfinding approach of inspection programs. In the TQC, with industry being responsible for the routine testing program, which is simply monitored by the FSIS, the program loses its criticism based character and becomes one which can be laudatory. Quality control by the industry merits congratulations and should be encouraged by all available means by the regulatory agency.[14]

Methods used for detection of *Salmonella* in processed foods frequently use combinations of old and or new technologies for the detection and identifications of these microorganisms. The specific combinations of the technologies vary in different methods, and the utility and efficiency of the methods is dependent upon the proper execution of these combinations. In general, methodologies for the analysis of the presence of *Salmonella* in foods are designed to improve sensitivity while reducing both cost of analysis, and cost to the company in terms of the time a product must be held up before sale and disposal, and time required to make the best use of the method.

To best understand the utility of methods which are likely to produce the best results for a particular plant in a particular industry, it is essential that the growth characteristics of *Salmonella* are thoroughly understood and appreciated. It is also essential that the user of analytical methods for microorganisms understand that certain characteristics of a particular strain of microorganism may not be stable under all conditions over time. These are, after all, living organisms, subject to great variation, and genetic change because they are single cells exposed to a variety of stresses and environments in their growth. While certain general rules apply to the growth of any microorganism, and *Salmonella* in particular, it must be understood that environmental factors which will affect the growth or survival of microorganisms are interdependent, and changes in one will affect the other and will affect the survival of the microorganism.[15]

In processing foods, the plant must keep controls and records, and therefore must have a system for identifying products and times of production. This system may ease the needs for testing or microbiological examination of unrealistic percentages of the finished products. For this purpose a "Production" lot has been defined as "the plant's designation . . . of all of the units of a product of the same size produced under essentially the same condition." The samples of this lot can be tested and all units of the lot can safely be assumed to essentially conform to the results obtained. Generally speaking, examination of 10 units per production lot would represent an adequate sampling of the product.

Since not all contamination of a food product will consist of *Salmonella*, it is also desirable to do plate counts to determine the total aerobic contamination of a food. To do so usually requires that the product be blended or otherwise thoroughly mixed in a sterile diluent (usually a buffered salt solution) for dilution to result in between 30 to 300 colonies per plate. This range in number of colonies permits an accurate count of the colonies and provides sufficient numbers to allow for desired accuracy. Based on the dilution used, the number of bacterial cells in 1 g of the original product can then be calculated.

In performing this plate count, it is essential that each sample be representative of the lot. Although 25-g samples are required to be used for *Salmonella* testing, so large a sample should not always be necessary for the plate count since the identification of one species or type of bacteria is not being sought here. Raw foods may well contain large numbers of mixed bacterial populations. The plate count seeks to determine the gross amount of contamination, and for some food types the numbers may be great. The sample selected should be completely homogenized and mixed in the first diluent tube or bottle, and in each dilution step thereafter. Small-sized samples (as low as 1 g) may be selected if a number of replicates (5 to 10) are done. If larger samples are used, fewer replicates are required.

From the first dilution to the final dilution, the sample must be mixed or blended in a standard manner. If hand mixing is used, the length of the arc of shaking and the number of arcs/unit time must be specified and must remain standard in order to assure that results can be compared and correlated. If mechanical mixing or blending is used, it should be done at a standard speed, for a standard time for the same reason. Differences in hand and mechanical mixing rarely permit results to be compared between the two methods.

Once mixing is accomplished of the final dilution(s), and frequently more than one is used to assure hitting the 30 to 300 range of colonies, 0.1 to 0.3 ml of the dilution is plated. Two different methods of plating may be used; one is a pour plate, the other a spread plate. For the pour plate method, Plate Count Agar is prepared, sterilized, and cooled in a water bath of 44 to 46°C. For convenience, it is best if the agar is sterilized in tubes in the quantity to be used (usually 15 ml), so that tubes can remain in the water bath until actually used. In this way the tube will remain at the correct molten temperature, neither too cool so as to begin to solidify and lump, nor too warm so as to destroy some heat sensitive bacteria. Using this method, multiple plates, usually three, can be prepared from each dilution by plating the dilution (0.1 to 0.3 ml each) into sterile petri dishes. Immediately, the 15 ml of molten agar is poured into the dish, and the entire dish is swirled to gently mix the sample throughout the agar. When the agar has solidified, the plates are inverted to avoid condensation of moisture on the agar surface with resulting smearing of colonies to produce confluent growth.

Inverted plates are incubated at 32 to 37°C for 48 ± 3 h prior to counting. The most common temperature selected for incubation is 35 ± 1°C. For some foods, particularly meats and dairy foods, the spoilage organisms are sometimes psychrophiles (growing at lower temperatures). For such food samples, either additional replicates, or all of the plates as desired may be incubated at 20 ± 1°C for a longer time: up to 5 days (120 ± 3 h). When this temperature and time is used, higher plate counts are often obtained.

The second type of plate count procedure is called the spread plate. In this case the plate count agar plates are poured, predried, and inverted for storage until use. The sample from the proper dilution is placed on the plate surface, either in 0.1 or 0.3 ml amount, and is immediately spread with an alcohol flamed glass hockey stick so that the sample is evenly spread over the entire surface of the plate. This technique takes some practice, and for plates to be

TABLE 8.1
Characteristics of Indicator Organisms

1. Applicable to all types of water
2. Present in sewage and polluted waters when pathogens are present
3. Number is correlated with the amount of pollution
4. Present in greater numbers than pathogens
5. No aftergrowth in water
6. Greater survival time than pathogens
7. Absent from unpolluted waters
8. Easily detected by simple laboratory tests in the shortest time consistent with accurate results
9. Has constant characteristics
10. Harmless to man and animals

From National Resource Council, *Drinking Water and Health*, National Academy of Sciences, Washington, D.C., 1977. With permission.

compared one to the other, the same technician should do the spreading because small differences in technique will make a great difference in the results obtained. The use of multiple plates for each dilution helps to minimize the effects of differences in spreading technics. One person, in doing these plates, should establish a routine, to be repeated on each plate; of where the hockey stick should first be placed and the direction and number of rotations to be used for each plate. Such a standard routine will help to minimize differences obtained in spreading the sample over the surface of the plate. Following spreading of all plates, the plates are again inverted, and incubated under the same conditions as described for the pour plates.

Plate counts will rarely be used to analyze the finished food product. Rather, it will be used to determine the quality of raw or partially processed foods which may have been contaminated. The results in these cases can be used to help the processor to determine the necessary limits of treatment or processing which will be required to achieve the desired quality of finished food product.

Water quality is measured on the basis of the presence or absence of a certain group of organisms used as indicators of fecal pollution (see Table 8.1). These coliform indicators have been used by some segments of the food industry to give some measure of the quality of, particularly, raw or partially treated food products. Some authorities question the need for coliform analysis of food products since they feel that these organisms have little significance in the processing of foods. The product in question, and the situation of each individual processor may determine whether testing for *Escherichia coli* is desired or used. When such testing is done, the procedures include presumptive, confirmative, and completed tests to determine the presence of the organisms and to definitely establish the identity of the bacteria detected.

Salmonella are facultative anaerobes whose growth is faster under aerobic conditions. These organisms will grow on a minimal medium of glucose as a carbon and energy source, ammonium salts as a nitrogen source, and some other mineral salts. Certain strains of some species do require special amino acids or vitamins, or sometime both. Most strains of *Salmonella* are easily killed at 60 to 65°C; however, this characteristic is altered by the water activity value of the medium.[15] Since this value is different for different foods, an average or absolute temperature for kill cannot be relied upon. The time required for 90% kill, at this temperature, of different *Salmonella* strains in different media has been reported to vary from 0.6 to 37 min. These reports make obvious the danger of assuming one time required for kill of these organisms in any food. Sublethal heating of *Salmonella* strains may result in cells which undergo repair and recovery to later increase in numbers and provide the infectious dose in foods consumed after this growth period. The effects of drying or freezing on *Salmonella* organisms has also been found to vary greatly from strain to strain and are heavily dependent upon the water activity of the growth medium.[17] The minimum water activity for growth of *Salmonella* is approximately 0.93 to 0.94.[18] Variations in water activity of different foods or media are a major factor in differential survival of various strains of *Salmonella* in these environments[17] and should be considered when determining times needed to reduce or eliminate contamination from a particular medium.

Radiation resistance appears to be controlled by a different mechanism than heat resistance in *Salmonella*. It has been shown that repeated doses of gamma radiation can result in development of resistance to radiation in certain strains of these bacteria. UV radiation appears to exhibit the same effects on the *Salmonella* as on other bacteria, but the pattern of sensitivity does not appear to be the same as heat sensitivity patterns in these organisms.[15]

Conventional schemes which have been in use for the isolation and identification of *Salmonella* from food samples have involved a series of steps, the completion of which is time-consuming, expensive, and requiring considerable expertise in the reading and interpretation of results. Such schemes involve pre-enrichment of the food sample in a liquid medium which will encourage the growth of any strain of *Salmonella*, followed by selective enrichment which will not only encourage the growth of these organisms, but will in addition inhibit the growth of other organisms which may be contaminating the sample. Following this selective enrichment, the organisms which grow should be plated to selective, and if possible differential media to encourage the growth of *Salmonella* strains and to allow ready identification or recognition of colonies which may be *Salmonella*. Growth on these media should then be read by trained personnel and suspect colonies should be picked to media which will permit the growth and positive identification of *Salmonella* strains by the biochemical reactions that occur after growth in these media. More positive identification can be made by use of serological typing with polyvalent, group, and specific antisera than with biochemical reactions only. In some extreme cases, phage typing can be used to identify *Salmonella*; however, this method is generally expensive and is not found necessary when the other methods are available.[15] Quantitating *Salmonella* in food samples is usually difficult because the numbers present are most likely to be small, and uniform sampling is most difficult to assure accuracy. The use of most probable number methods for estimation of the number of living cells present is useful, providing the sample is prepared in such a way as to assure even distribution of organisms present. A complete discussion, including tables for use, can be found in several references, including American Public Health Association (APHA)[19] and others.[20,21] The most probable number (MPN) is a method to establish an estimate of the number of viable organisms present in the sample under consideration. In this respect, it is different from the direct microscopic count which enumerates both living and nonliving cells; and it is similar to the agar plate count which enumerates only living organisms (most often now referred to as colony forming units). To obtain an MPN, the theory of probability must be applied to test results, with a number of assumptions taken as given. For samples in which the number of viable cells is high, the MPN is not as accurate as the agar plate count. The MPN is always an estimate — not an experimentally or culturally determined number. Depending upon the accuracy, or confidence limits desired, the MPN may be determined by 3, 5, or 10 tube tests.

Recently, a number of newer methods have been introduced, generally with the aim of reducing the time required for the tests, and therefore, from the standpoint of the food processor, the cost of maintaining this quality control program. Time required for testing is only a portion of the time and cost which may face the processor. If a product is found to contain *Salmonella* then the entire contaminated lot must be quarantined by the processor, increasing the cost to this manufacturer. Many of these methods have continued to be based on the fermentation patterns of the different strains of *Salmonella*, but some have involved other, more stable, and more rapidly demonstrable characteristics.[15] In some of the methods of this type which have recently been introduced, the method consists of nothing more than a "kit" which contains the required battery of fermentation media needed for identification. Such "kits" rarely contain a means for determining the concentration of *Salmonella*, and some method of estimation such as that described above is required. Negative tests are easily interpreted as meaning that *Salmonella* are not present; however, suspect samples must be confirmed by conventional methods which may require longer times.

In addition to the conventional fermentation reactions used for identification of suspect organisms, fluorescent antibody tests of several types have been described. The FDA Manual[21] lists a fluorescent antibody method for use as a screening method for the presence of *Salmonella* which has been approved for this purpose. It is specified that since the fluorescent tagged antibodies cross-react with other members of the Enterobacteriaceae, these tests must be confirmed by the traditional cultural methods approved in the manual. These tests have the advantages of speed and the reliability of antibody specificity when properly designed and executed. The methods utilizing fluorescent tagged antibody are generally limited by the reagents used and the skill and training of the person performing the test. The preparation of the reagents is critically important to the reaction in a test. A great advantage to the use of fluorescent antibody methods in any identification scheme is that they may be used for such a wide variety of organisms and have been used for the *Salmonella* and many of the closely related Enterobacteraceae.[22-25]

Other techniques which have been tried for the rapid identification and enumeration of *Salmonella* contaminants in food products have included specialized chromatographic procedures, enzyme-tagged antibody, membrane disc techniques, DNA composition (guanadine-cytosine ratio), and combinations of these methods. All procedures used for detecting *Salmonella* in foods must be compared statistically with the Official AOAC Method[3] before being accepted for official use. Official use of these newer methods is generally limited to use as rapid screening methods and may be used for large numbers of food samples and should be useful in quality control programs, but these must be confirmed by more conventional cultural and serological methods when suspect samples are encountered.

A rapid method involving several cultural, concentration, and serological techniques is that recently described as the "Salmonella Bio-enzabead™ Test Kit" which has been approved by the AOAC[25] as a screening method which can be used with large numbers of food samples. This specific kit is an example of a group of tests known as enzyme immunoassays which are based on the linkage of an antibody or antigen to an enzyme without greatly affecting the activity of each. This type of system is easy to prepare, is relatively inexpensive in consideration of all the different media used in standard methodology, and has more sensitivity than some of the older, routine methods. One requirement, however, is that the antibody used must be specific for the antigen expected in the food, or the results will be confused by false positive reactions. In the *Salmonella* groups, this requirement can result in some limitations if the strains of *Salmonella* being sought and expected are varied in one location or for one food processing plant. In the Bio-enzabead™ test, the procedure can be used only as a screening method because some non-*Salmonella* antigens cross-react with the extremely specific monoclonal antibody used in the test. When reactions are observed in the screening, confirmation of the *Salmonella* nature of the culture must be made by the conventional tests previously described for the identification of these organisms.[6-8,] Similarly, the 1-2 test™ detects motile *Salmonella* by the formation of immunodiffusion bands in an agar medium.[26] Indication of positives must still be confirmed by conventional cultural methods.

Technological developments which promise utility in rapid screening for the presence of contaminant organisms such as *Salmonella* include the use of DNA probes, which can be used to identify the genetic information of any organism, including contaminating bacteria.[9,13,27] These bioprobes are chemically stable and kits have been developed to detect and identify specific contaminations or infections. Such a method which is used in some laboratories as a qualitative means of detection of *Salmonella* species in foods is that developed by the Gene-Trak Systems.[27] This method uses *Salmonella*-specific, DNA probes and a colorimetric detection system for *Salmonella* species in food samples after those samples have been enriched in broth culture. These tests are based on the knowledge that most bacterial organisms like the *Salmonella* possess DNA comprised of two complementary strands. In the test, the DNA is released from the organism of interest into the medium, and then denatured to become a single strand. Following this a probe which is labeled so that the resulting hybrid DNA molecules can be identified is placed into the medium to react with the complementary single strand from the objective organism. If the probe used is identical to that needed for the *Salmonella* strain, when the reaction occurs, the single strand DNA is detected. The Gene-Trak system comes with detailed methods for analysis.[27]

Monoclonal antibodies, as used in the Bio-enzabead™ test,[25] can be produced in large quantities which are homogeneous and of monospecificity, are extremely useful in detecting and identifying contaminating organisms[16] in a variety of different tests. Such techniques must be properly adapted to use in specific instances such as screening large numbers of food samples. The adaptation for use in these circumstances is critical to the utility of the methods involving such techniques, and successful adaptation of monoclonal antibodies has been invoked in the Bio-enzabead™ system.[25] Some specialized tests systems are approved for certain food types. These systems may be exemplified by the hydrophobic, grid-membrane filter method for rapid detection of *Salmonella* in chocolate, raw poultry, pepper, powdered cheese, powdered egg, and in nonfat dry milk. This method requires special hydrophobic filters as specified, and the necessary equipment to accomplish filtration. The system approved recommends the use of materials provided by QA Laboratories in Toronto, Ontario, Canada. The different foods are treated according to specifications for each and are inoculated into selective broths for short incubation times. Finally, the filtration of the selective broths through the hydrophobic filters is followed by placing the filters on the surface of predried Selenite lysine agar (SLA) or Hektoen (HE) enteric agar plates for incubation for 24 h (SLA at $43 \pm 0.5°C$, and HE at 35°C). On SLA, *Salmonella* appear as blue to purple colonies, indicating a lysine positive reaction. Other colonies will be yellow to yellow-green. On HE, *Salmonella* will produce blue colonies with black centers, with some strains producing colonies which are almost completely black. Typical or suspicious colonies are picked to Triple Sugar Iron (TSI) agar slants, Lysine Iron Agar (LIA) and MacConkey's (MAC) agar and incubated for 24 h at 35°C. The specific biochemical and serological identifications then must follow this procedure.

If approved by the FDA and the AOAC, the primary considerations in selection of a method are the expense of the reagents, the time required for completion of the method when samples are positive for the presence of *Salmonella*, and the comparison of the efficiency and specificity of the method with the methods recommended by the AOAC. The primary reason for change of methods in the past has been improvement in the time required and/or the specificity obtained. Time required to complete the test determines how long food products must be held before distribution and sale, and for all practical purposes this determines the cost of the method in relation to product sale.

Organisms, including *Salmonella* contaminants present in food samples, have often been subjected to some stress in some form of food processing. It is, therefore, essential that the prepared food sample first be subjected to an enrichment period, including a nutritious, noninhibitory medium in order that the organism may recover from possible injury or shock, and be restored to a stable state. The first step in testing is always preparation of the food sample for isolation. In general, the tests approved by the FDA depend on the use of 25 g samples of the food at a ratio of 1:9 (sample to broth). This size sample provides sufficient material so that contamination which is present will be detected, while the ratio of sample to broth provides sufficient liquidity and mixing so that any contamination present will be detected in the test. For some tests, the broth-to-sample composite is mixed with inhibitory reagents (i.e., brilliant green water) and incubated for up to 24 h without mixing or pH adjustment. This is deemed necessary for the recovery of injured or shocked cells and for inhibition of growth of bacteria other than *Salmonella* which may have contaminated the food sample. The incubation temperature of these samples is 35°C.

When whole animal carcasses are sampled for culture, as with chicken or rabbit meat, although the process is not actually spelled out by the FDA methods, it is essential that the laboratory use adequate amounts of liquid for washing the carcass to obtain a representative sampling of the surface of the meat — including the surfaces of the body cavities. In such cases, it is fairly common practice to use 250 ml of a buffered saline solution to wash the entire carcass. This wash is accomplished in a sterile plastic bag and the buffered saline can then be inoculated to the desired enrichment media in adequate amounts.

Following this preparation of the food, selective enrichment is performed on the incubated samples. The incubated samples are mixed by gentle shaking. Following mixing, 0.1 ml of sample is transferred into 10 ml of tetrathionate broth at a temperature of 25 to 35°C. The inoculated broth is then mixed to disperse the inoculum and incubated in a water bath at 35°C for 6 to 8 h.

A. FOOD SAMPLE PREPARATION FOR *SALMONELLA* ISOLATION

Most food samples are analyzed on the exact weight basis, that is the 25-g samples. Some, such as frog legs, require special methods for preparation. Some authors recommend two 25-g samples from each food to be tested, each weighed into separate sterile containers.[20]

In general, foods are grouped together on the basis of constituent content for compositing in preparation for culture for the presence of *Salmonella*. The following are specific methods for groups of foods.

1. Foods including dried egg (yolks or whites), dried whole eggs, pasteurized liquid and frozen eggs, prepared dried and powdered mixes (cake and other bread and pastries), infant formula, and nutrient formulas for oral or tube feedings are aseptically weighed into 25-g samples into sterile wide-mouthed jars with screw-caps. The containers should have a capacity of at least 500 ml. If the sample is a powdered sample, 15 ml sterile lactose broth is added and stirred until the suspension is smooth. Additional lactose broth is added in the amounts of 10 ml, 25 ml, and 190 ml to total 225 ml, and stirred thoroughly until the sample is smoothly suspended without lumps. This container is closed and allowed to stand for 60 min at room temperature. The pH is adjusted to 6.8 with sterile 1 N sodium hydroxide or 1 N hydrochloric acid; the closure is loosened for gas exchange, and the sample is incubated for 24 h at 35°C. If the sample is nonpowdered, the 225 ml of sterile lactose broth is added at one time, and the mixing and pH adjustment are the same as in the powdered sample. If a sample must be thawed from the frozen state, carefully follow instructions for thawing found in the *FDA Bacteriological Analytical Manual*.[21] After incubation, the culture procedure continues as described below.

2. To prepare samples of egg-containing products such as noodles, egg rolls, macaroni, spaghetti or other pasta, cheese, dough, prepared salads (such as ham, chicken, egg, tuna, turkey, or pasta), dried, fresh, or frozen fruits and vegetables, nut meats, shrimp, crab, crayfish, langostinos, lobster, and fish, do not thaw frozen samples if possible before analysis. If frozen samples must be thawed in order to obtain the analytical sample, then the instructions for thawing must be followed exactly (i.e., either thaw below 45°C for 15 min with continuous mixing in a thermostatically controlled water bath, or within 18 h at 2 to 5°C. Sterile lactose broth (225 ml) should be blended for 2 min in a sterile blending container with the sample, then transferred to a capped jar of at least 500 ml capacity to stand for 60 min at room temperature before pH adjustment to 6.8 and incubation for 24 h at 35°C prior to cultures.

2a. For egg products which are not pasteurized, and are frozen, the samples should not be thawed before analysis. If it is necessary to thaw the frozen sample to obtain the correct sample for analysis, thaw below 45°C for about 15 min with constant agitation, or thaw within 18 h at a temperature of 2 to 5°C. The 25-g sample is then enriched in 225 ml of selenite cystine broth in one container, and a 25-g sample is enriched in 225 ml tetrathionate broth plus 2.25 ml of 0.1% brilliant green dye solution for 24 h at 35°C.

3.   For samples of meats, meat substitutes, meat by-products, animal substances, glandular products, and fish, meat, or bone meals, the 25-g samples should be weighed into a sterile blending container to which is added 225 ml sterile lactose broth. Bacteria present in these samples may have been subjected to freezing, drying, or chemical treatment which may have left them in a seemingly attenuated state, and the use of lactose broth for these samples may help bring the bacteria back to a more normal state. Lactose is not a sugar usually fermented by the *Salmonella*; however, the fermentation of this sugar by other contaminating organisms may lower the pH to a point of inhibition of competing bacteria, while the *Salmonella* are less affected, and will continue to grow. The composite is blended for 2 min, and then transferred to a capped jar of at least a 500-ml capacity to stand for 60 min at room temperature. Following this, the pH should be adjusted to 6.8 followed by the addition of 2.25 ml of Tergitol Anionic 7 which has been steamed for 15 min. The composite should now be mixed well. If Tergitol Anionic 7 cannot be used, Triton X-100, also steamed for 15 min should be used. In any case, the amount of these surfactants should be kept to a minimum to permit foaming. The mixed composites should be incubated for 24 h at 35°C before cultures are continued. When the samples of the meat products above are raw or highly contaminated, duplicate 25-g samples should be weighed into separate sterile blending containers. In one blender, selenite cystine broth (225 ml) should be added, and to the other 225 ml of tetrathionate broth without brilliant green. Blending occurs for 2 min in both containers before the samples are transferred to separate, 500-ml containers to stand for 60 min at room temperature. The pH should be adjusted to 6.8, following which 2.25 ml of 0.1% brilliant green dye solution is added to the sample enriched in tetrathionate broth. The samples should be mixed by swirling, and incubated at 35°C for 24 h before cultures are continued.

4.   Coconut samples must also be mixed during processing with surfactant to initiate foaming. The 25-g sample of coconut should be weighed into a capped 500-ml-capacity container to which is added 225 ml of sterile lactose broth. The composite is allowed to stand for 60 min at room temperature, mixed, and the pH tested prior to adjustment to pH 6.8. Following pH adjustment the minimum quantity of steamed Tergitol Anionic 7 or steamed Triton X-100 is added to initiate foaming. The

sample is then mixed and allowed to incubate for 24 h at 35°C before cultures continue.

5. For testing samples of gelatin, the 25-g sample is weighed aseptically into a sterile capped jar of at least 500-ml-capacity to which is added 225 ml of sterile lactose broth and 5 ml of 5% aqueous gelatinase solution. The composite is mixed well and allowed to stand at room temperature for 60 min before adjusting the pH to 6.8 and incubating for 24 h at 35°C before continued cultures are performed.

6. Casein samples of 25-g are weighed aseptically into sterile blenders, and then blended for 2 min with sterile lactose broth. Following blending and standing at room temperature for 60 min, the mixture is transferred to a sterile 500-ml-capacity capped jar; pH is adjusted to 6.8, and the mixture is incubated for 24 h at 35°C prior to further culture.

7. Soy flour samples may not be composited as other samples and 25-g samples should be aseptically weighed into a sterile beaker. Then, using a sterile funnel, the sample is poured gently and slowly over the surface of 225 ml of sterile lactose broth in a 500-ml-capped flask or jar. Without mixing or pH adjustment, the mixture is incubated for 24 h at 35°C before further culture.

8. Rabbit meat is a food which cannot be analyzed in the usual 25-g sample. Rather, the entire carcass is sampled. Three carcasses are placed in a sterile plastic bag and covered with sterile lactose broth. The filled bag is placed in a suitable container and shaken for 15 min on a mechanical shaker (100 strokes per minute). Following this, the lactose broth is removed into another sterile container, and additional broth is added to a total volume of 3500 ml. This broth sample is allowed to stand for 60 min at room temperature before adjusting the pH to 6.8. The broth is then incubated for 24 h at 35°C prior to further culture.

9. Frog legs are also a food which cannot be analyzed in the usual 25-g sample. Rather, 15 pairs of frog legs are placed into a sterile plastic bag, as in the case of the rabbit carcasses, and covered with sterile lactose broth. The bag is placed into a large container and placed on a mechanical shaker (100 strokes per minute) for 15 min. Following the shaker, the broth is poured into another sterile container, and the volume of broth is made up to 3500 ml by the addition of sterile lactose broth. After mixing, the broth is allowed to stand at room temperature for 60 min before the pH is adjusted to 6.8, and the container of broth is incubated at 35°C for 24 h prior to additional culture.

10. For the examination of dry, whole milk, a 25-g sample is weighed aseptically into a sterile capped jar of 500-ml capacity to which is added 225 ml of sterile distilled water. The composite is mixed by swirling and allowed to stand at room temperature for 60 min. The pH is determined and adjusted to 6.8 before the addition of 0.45 ml of 1% brilliant green

dye solution. After mixing, the container is incubated for 24 h at 35°C before additional culture.

11.   Other foods which must use nonselective diluents include candy and chocolate candy coatings. To the 25-g sample, 225 ml of sterile reconstituted dry nonfat milk is added and the sample is blended for 2 min in a sterile blender container. The mixture is aseptically transferred to a sterile capped jar of at least 500-ml capacity and allowed to stand at room temperature for 60 min. After mixing, the pH is determined and adjusted to 6.8, followed by the addition of 0.45 ml of 1% brilliant green dye solution. The composite is then incubated for 24 h at 35°C prior to additional cultures.

12.   Frosting and topping mixtures are aseptically weighed into 25-g samples in sterile capped jars of 500-ml capacity. To the sample is added 225 ml of sterile nutrient broth. The composite is mixed well and allowed to stand at room temperature for 60 min prior to adjusting the pH to 6.8. The composite is then incubated for 24 h at 35°C prior to further testing.

13.   Food dyes and food colorings must be analyzed by methods based upon the pH of the aqueous suspension of the material. For products with a pH of 6.0 or above, the same method is used as for dried whole egg sample preparation. For those materials with a pH below 6.0, the 25-g sample is aseptically weighed into a sterile capped jar of 500-ml capacity to which is added 225 ml of tetrathionate broth without brilliant green dye. The sample and broth are mixed and allowed to stand for 60 min at room temperature prior to adjustment of pH to 6.8. After the pH is adjusted, 2.25 ml of 0.1% brilliant green dye is added to the sample and mixed thoroughly prior to incubation for 24 h at 35°C.

14.   Because of mixing properties, nonfat dry milk, instant and noninstant are handled somewhat differently for sample preparation. The 25-g sample of noninstant milk cannot be composited, whereas the sample of instant milk may be. Using a sterile funnel, the 25-g sample is aseptically weighed into a sterile beaker and slowly poured over the surface of 225 ml of brilliant green water in a sterile 500-ml-capacity container. The brilliant green water is prepared by the addition of 2 ml of 1% brilliant green dye solution to 1000 ml of sterile distilled water. After pouring the milk sample gently over the surface of the brilliant green water, the composite should stand for 60 min at room temperature without disturbance. Without adjusting the pH, the sample is then incubated for 24 h at 35°C.

15.   Dried yeast samples are aseptically weighed in the 25-g amounts into sterile capped jars of 500-ml capacity to which is added 225 ml sterile trypticase soy broth. After mixing well, the sample is allowed to stand for 60 min at room temperature before adjusting the pH to 6.8. The sample is then incubated for 24 h at 35°C. If the sample is active yeast, the incubated sample is mixed and 1.0 ml is transferred to each of 10

ml lauryl tryptose broth and 10 ml of tetrathionate broth. These broths are incubated for 24 h at 35°C, and vortexed and streaked to 3 selective agars (HE, Bismuth sulfite [BS], and Xylose lysine desoxycholate [XLD]).

16. The examination of spices of various kinds present several special situations which must be adjusted for in the preparation of samples for analysis. Generally speaking, black pepper, white pepper, celery seeds or flakes, chili powder, cumin, paprika, parsley flakes, rosemary, sesame seed, thyme, and other vegetable flakes may be analyzed by the same method. This consists of aseptically weighing the 25-g sample into a sterile capped container of at least 500-ml capacity to which is added 225 ml of trypticase soy broth. The sample and broth is mixed and allowed to stand at room temperature for 60 min before the pH is adjusted to 6.8. Prior to further culture, the sample is then incubated at 35°C for 24 h.

Onion and garlic flakes and powder, because of certain chemical compositions, must be treated somewhat differently in that the 25-g sample is pre-enriched in trypticase soy broth to which has been added 0.5% potassium sulfate (the addition is made prior to autoclaving the 225 ml sample of trypticase soy broth). After this addition, the sample mixture is allowed to stand for 60 min at room temperature, prior to adjustment of the pH to 6.8, and incubation for 24 h at 35°C.

Allspice, cinnamon, cloves, and oregano all contain some toxicity for which there is no known neutralization. These spices, therefore, must be examined at a dilution above the toxic level, which is 1:100 for allspice, cinnamon, and oregano, and 1:1000 for cloves. For heavy condiments, a sample broth ratio of more than 1:10 must be used because the dehydrated condition of the condiment absorbs so much of the broth. Using the proper sample to broth ratio for the spice being examined, the procedure is then the same as that described for the peppers.

The sample preparations and incubations described here conclude the pre-enrichment step for the isolation and identification of *Salmonella* from foods. The next procedures are included in the selective enrichment steps which allow a continued increase in the numbers of *Salmonella* while at the same time inhibiting the increase of other bacteria which are likely to be present in or on the food samples.

In general, foods which must be examined will contain only small numbers of *Salmonella*, if any, and therefore other types of bacteria may well outnumber them greatly. The selective enrichment steps to be used, then, must be directed toward allowing the *Salmonella* to increase in numbers while inhibiting other bacterial genera and preventing their growth. The culture media used were almost universally designed to isolate *Salmonella* from other bacteria when the mixtures were present in feces. That situation is very different from the environment, and the mixtures which are to be found in most foods, and it is somewhat surprising that the enrichment media work as well as they do when used in attempts at isolation of these organisms from food samples.

The two enrichment broths most recommended by the USDA are tetrathionate broth (including the modifications as have been recommended in several cases) and selenite broth. There are other broths which will serve to enrich *Salmonella* numbers in mixtures of bacteria, including GN Broth and Hajna, although this medium has been reported to work better for the enrichment of *Shigella* species than for *Salmonella*. The purpose of selective enrichment is to provide an opportunity for the small numbers of *Salmonella* likely to be present to grow and increase in numbers while the competing organisms which are likely to be present in greater numbers are inhibited. The inclusion of selenite and thiosulfate in the first two enrichments seem to provide a better selection for the *Salmonella* organisms than do the other selective media, while the buffering action of the calcium carbonate in the tetrathionate prevents a great change in pH which may be detrimental to *Salmonella*. The tetrathionate present may not be completely harmless to *Salmonella*, but will inhibit the growth of other microorganisms to a greater extent since there is more toxicity for other bacteria than for the *Salmonella*. One modification of tetrathionate, TT Broth, has been greatly enriched by the addition of yeast extract which allows more rapid growth of many bacteria.

A pre-enrichment step and culture is sometimes recommended for the recovery and identification of microorganisms which may have suffered some injury or stress in some operational step of a process. In this situation, the food sample should be mixed with a nonselective broth prior to the use of selective media and isolation steps. This permits the recovery of the organisms from any stress which may have occurred during subjection to heat, dessication, preservative action, osmotic pressure, or pH changes in the foods prior to sampling. The pre-enrichment process was devised for use with certain types of foods and permits dilution of substances present in the food which may be inhibitory to the growth of the bacteria. Several different pre-enrichment broths and procedures have been recommended in different situations; frequently a lactose broth. APHA reports that in some cases there may be an unfavorable coliform-salmonella ratio which will interfere with the recovery of the salmonella, and when this occurs the use of selenite broth containing 10% sterile feces improves that recovery.[1] Some researchers have recommended different pre-enrichment broths for different food samples and have also reported improved salmonella recovery when an incubation temperature of 43°C is used rather than the usual 37°C. The most common pre-enrichment broths recommended are lactose and brilliant green, although lauryl tryptose broth, mannitol purple sugar broth, and nutrient broth have been recommended.

The enrichment effect of selenite broth, which also has a peptone base as does tetrathionate, is said to be due to the inclusion of a carbohydrate which, when fermented, results in a lower pH thereby inhibiting the growth of Enterobacteriaceae. The addition of cystine to one modification of selenite broth is seen to enhance the growth of *Salmonella* cells in comparison to other bacteria when large amounts of organic materials are present.

The use of selective enrichment broths, while allowing the development of large numbers of *Salmonella*, also result in mixed cultures with large numbers of other organisms as well. It is therefore, necessary to follow selective enrichment procedures with plating to highly selective media which will now allow the development of discrete *Salmonella* colonies, while hopefully inhibiting the growth of colonies of other bacteria. Growth of discrete colonies on these media also permits recognition of colonies which are suspected to be *Salmonella* and permits transfer of these pure colonies to other media for confirmation of identity.

Selective and differential media which can be used in this step of the procedure include Salmonella-Shigella (SS) agar, BS agar, and XLD agar. Others which could also be used are Brilliant Green agar and Desoxycholate citrate agar. Those media which are less selective but primarily differential are MAC agar, HE agar, Desoxycholate agar, and Eosin-methylene Blue (EMB) agar. The inclusion of the primarily differential media provides a better opportunity to obtain growth from all isolates, even those which are most likely to be inhibited by the highly selective media.

SS agar cultures of *Salmonella* should result in colorless colonies on the pink agar background. Many enterobacteria colonies, not members of the *Salmonella* or *Shigella* genera, will appear as red colonies or will appear against a yellow background resulting from the fermentation of the lactose present in the medium. SS agar is very hygroscopic, and after a bottle is opened, it will rather quickly solidify, and frequently will then become highly toxic to microbial growth. Because of this the medium should be ordered in small containers and should not be retained in the laboratory for long periods of time. Medium which has solidified should generally be considered to be likely to be toxic and should be discarded in favor of freshly opened containers of the dehydrated medium. Those *Salmonella* strains which produce hydrogen sulfide will generally produce black centered colonies because of the inclusion of ferric and sodium sulfates and sodium thiosulfate as the indicator system for this metabolite.

BS agar does not depend upon fermentation by the bacterial colonies for recognition of type. Rather, the inhibitory nature of the ingredients simply inhibit the growth of most Enterobacteriaceae. The inclusion of the indicator system for the presence of hydrogen sulfide results in black colonies surrounded by a zone which will appear brown to black with a metallic sheen, similar to that seen on some media on the surface of *E. coli* colonies. To best produce the typical identifying reaction, plates of BS agar should be prepared at least 2 days before use. Drying which occurs in this time appears to enhance the typical appearance of the colonies.

Xylose lysine (XL) agar was originally developed for the isolation and identification of *Shigella*, but it has also been found to be useful in the selection and identification of *Salmonella*. In addition to the hydrogen sulfide indicator system, it also contains three sugars which may be fermented by some bacteria. The sugars are lactose, sucrose, and xylose. The agar also contains lysine, which is utilized by *Salmonella*, and after the xylose in the medium is exhausted by the *Salmonella*, they utilize the lysine, causing a reversion of the acid reaction to alkaline and a red color with the phenol red indicator. This reversion is not seen with coliforms because they have also fermented the lactose and the sucrose with a production of an excess of acid which cannot be neutralized and reversed. After 24 h incubation of plates containing *Salmonella* from enrichment, *Salmonella* colonies are typically yellow with black centers. In another 24 h the colonies turn from yellow to pink. The XL medium has been modified to XLD agar by the inclusion of sodium desoxycholate which prevents swarming or spreading of *Proteus* colonies when these bacteria are present in the sample being tested.

MAC agar is differential, but is less selective than the other media discussed above. The medium is a peptone-based agar with a small amount of bile salts and crystal violet included to inhibit at least some of the gram-positive organisms which may be present in the contaminated sample. Lactose is present in the medium as a carbohydrate which can be fermented by some organisms, but not by the *Salmonella*. When organisms do ferment lactose, they appear as red colonies against a background of precipitated bile, while *Salmonella* are colorless colonies against an unchanged background. The medium can be made more differential by the addition of sucrose to distinguish organisms which ferment that carbohydrate.

BGA is a peptone-based medium to which the dye has been added to suppress the growth of gram-positive bacteria, and to which sulfapyratide can be added to suppress the growth of *Proteus*, a bacterium which is so commonly found to contaminate the same environments contaminated by *Salmonella*. The selectivity of BGA consists of two fermentable carbohydrates (lactose and sucrose) and an acid indicator, phenol red. Fermentation of the carbohydrates by other organisms results in the production of large amounts of acid which causes a yellow color, the *Salmonella* can be easily distinguished because of the production of alkaline products from the metabolism of peptone. Colonies of *Salmonella* can be easily distinguished because of the production of alkaline products from the metabolism of peptone. Colonies of *Salmonella* are pink (resulting from the phenol red in the medium) and are surrounded by a red zone in the medium. Care must be taken when testing a heavily contaminated sample because the production of many green or yellow colonies with the resulting acid in the medium may cause the *Salmonella* colonies to appear somewhat brownish with no red discoloration of the surrounding medium.

The use of lactose in so many of the selective and differential media causes one complication in the selection and picking of suspect colonies to identification media. There are a few nonconforming colonies of *Salmonella* which will ferment lactose, but when this does occur the fermentation is most often slow

— that is, up to 72 to 96 h may be required for the fermentation to occur. Rarely, a colony may ferment the carbohydrate more rapidly, and when this occurs, the organism cannot be easily distinguished from coliforms which cause this reaction routinely. These organisms have been the subject of much debate as to proper classification, but are generally classified now as *Salmonella* sp, subgenus III (Arizona) for the purpose of testing and identification of contaminants of foods. The organisms do infect the human, causing much the same symptoms, and derive from the same group of sources, as do other *Salmonella* organisms.

After plating to the selective media, regardless of the particular combination of media used, the plates are incubated at 35°C. After 24 h, BGA, SS, MCA, and XL plates are examined for the presence of suspicious colonies. If none are observed, the plates should be returned to the incubator for an additional 24 h and then should be reexamined when the BS plates are first observed, i.e., after a total of 48 h incubation. When suspect colonies are observed, they should be marked. At least three colonies of the same type should be selected at this time if there are many present, but if none are typical, at least three or four possible aberrant forms should be selected to be picked to screening media. All plates should be retained for reference if needed.

At least two rapid screening media should be used. When this is done, tentative confirmation of either positive or negative results can usually be obtained within 24 h. The two media recommended are TSI and LIA. These are used as slants in tubes for rapid reading of results to indicate the presence or absence of *Salmonella* on the selective plates. TSI contains small amounts of glucose and larger amounts of both sucrose and lactose. It has an indicator for hydrogen sulfide in the form of ferrous sulfate and thiosulfate, as well as phenol red as indicator for acid or alkaline reactions. The amount of glucose is sufficient to cause a yellow or acid reaction in the base or butt of the tube, but the acid is oxidized on the surface, which in reaction with the alkaline products from peptone metabolism cause the medium to turn red. If either sucrose or lactose is fermented, the slant will also be yellow because of the excess of acid. Reactions of *Salmonella* isolates will all show yellow reactions in the butt of the tube and red on the slant (except for those members of subgenus III which also ferment lactose and which will have a yellow slant). All subgenera III organisms will cause the production of hydrogen sulfide which will aid in their identification. The only species which will not form hydrogen sulfide are *S. pullorum*, *S. gallinarum*, and *S. typhisuis*. These species will show a yellow or acid butt and a red or alkaline slant.

LIA contains the same carbohydrate and hydrogen sulfide indicators which are present in TSI. However, LIA also contains the amino acid, lysine, which is decarboxylated by *Salmonella* strains, removing carbon dioxide. When this occurs, the reaction becomes alkaline due to the presence of the amine, and the cultures are, therefore, purple, rather than the usual yellow or red observed on other media. The production of hydrogen sulfide causes a black reaction

TABLE 8.2
Criteria for Discarding Non-Salmonella Cultures

Test or Substrate	Result
1. Urease	Positive (purple-red color)
2. Indole	Positive (violet color at surface)
Polyvalent flagellar (H) test	
or Spicer-Edwards flagellar test	Negative (no agglutination)
3. Lysine decarboxylase	Negative (yellow color)
KCN Broth	Positive (growth)
4. Phenol Red Lactose Broth	Positive (yellow color and/or gas)[a,b]
5. Phenol Red Sucrose Broth	Positive (yellow color and/or gas)[b]
6. KCN Broth	Positive (growth)
Voges-Proskauer Test	Positive (pink-to-red color)
Methyl Red Test	Negative (diffuse yellow color)

[a]   Test malonate broth positive cultures further to determine if they are *Salmonella arizonae*.

[b]   Do not discard positive broth cultures if corresponding LIA cultures give typical *Salmonella* reactions; test further to determine if they are *Salmonella* species.

From Food and Drug Administration, *Bacteriological Analytical Manual*, 6th ed., Association of Official Analytical Chemists, Arlington, VA, 1984, 7, 15.

on the medium which is indicative of *Salmonella* species when observed in conjunction with other characteristics of the growth on this medium. Reactions of most *Salmonella* on LIA include a purple butt with the production of hydrogen sulfide apparent by all species except *S. pullorum*, *S. gallinarum*, and *S. typhisuis*.

If biochemical testing is to be fully utilized to confirm a colony as *Salmonella* or to assign it to another genus, then the minimum number of biochemical tests must be done. These generally are considered to be those included in the discussions of *Salmonella* identification in the *FDA Bacteriological Analytical Manual*.[21] These include glucose (which can be read from the TSI slant), lysine decarboxylase (read from the LIA slant), hydrogen sulfide (read from either of the above slants), urease, indole, Voges-Proskauer, citrate, methyl red, motility, failure to ferment sucrose or lactose (few exceptions in the case of lactose), and growth in KCN broth.

The *FDA Bacteriological Analytical Manual*[21] lists the criteria which must be considered before any assumed non-*Salmonella* cultures are discarded. If cultures to be tested have been examined against these tests, and the reactions have been confirmed, then serological testing can be continued, and the species and type confirmed by those tests. Criteria for discarding are tabulated in Table 8.2.

The *USDA Microbiology Laboratory Guidebook*[2] (currently being revised) states that the tests needed to establish that a culture is within the family Enterobacteriaceae (whether to confirm it as a *Salmonella* or to place it in another genus) include production of indole, a negative methyl-red reaction, production of acetyl-methyl-carbinol from glucose (a positive Voges-Proskauer test), growth on citrate medium, hydrolysis of urea, liquefaction of gelatin, growth in KCN, presence of decarboxylase enzymes for lysine and ornithine, arginine dihydrolase, phenylalanine deaminase, ortho nitrophenyl-B-galactosidase, motility, no gas production from the fermentation of glucose, no acid production or fermentation of dulcitol, salicin, mannitol, lactose, sucrose, maltose, adonitol, sorbitol, inositol, arabinose, raffinose, rhamnose, trehalose, malonate, mucate, and d-tartrate. It is not necessary that all of these tests be done each time a culture is tested. When it is established that a culture is *Salmonella*, no further tests need be done before serological testing is begun. Because of the expense involved, it is preferable that serological testing follow biochemical testing, not precede it.

As mentioned previously, some media, particularly those which are selective become toxic and will not grow even the most hardy and nonselective cultures. It is, therefore, necessary at times to evaluate plating media which are to be used in the testing of food samples. Any medium is unsatisfactory for use if its efficiency in growing *S. typhimurium* is less than 75% when compared to trypticase soy agar. *S. choleraesuis*, *S. typhisuis*, *S. pullorum*, and *S. gallinarum* should not be used to establish expectations as concerns the more fastidious strains which may be studied on the medium in question. To do the efficiency test, use three plates of the medium in question and three plates of trypticase soy agar which have been previously dried following pouring. Exactly 0.1 ml of inoculum from each culture prepared in Butterfield's buffered diluent is used for spreading on the surface of the plates with a glass hockey stick spreader. Plating should be completed within 15 min of the preparation of the dilution. The plates are incubated at 35°C for 48 h before colonies are counted. At that time the number of colonies counted on the plating medium is divided by the number of colonies counted on the trypticase soy plates to determine the percent efficiency. While it is not necessary to test each bottle of dehydrated culture medium used for this work in this way, it would be advisable to test each new manufacturers lot of dehydrated culture medium in this way to determine what may be expected in routine work.

It is frequently necessary that laboratories test many samples which can obviously be expected to be negative for the presence of *Salmonella*. When this is necessary, much time, labor, and material may be saved if the samples are pooled at the nonselective enrichment step. This same pooling procedure can be used when testing many 25-g samples of a single food from a single source. It is most important that the volume of broth to sample be maintained as in the required procedure outlined previously. The USDA also makes certain recommendations which will assure accurate results when followed during pooling of samples for test. For example, when samples are pooled in the blenders, prior to culture in lactose broth, the broth should be warmed in advance to being mixed with the samples in blenders or culture flasks. The addition of warmed broth prevents additional stress being applied to organisms which may have been stressed in the food sample by various processes. When food homogenates are to be tested as a pool, the temperature should be brought to 35°C before the cultures are transferred to the incubator. If very large pools are to be tested, the incubation should be prolonged to 2 days, and subcultures of 10 ml of the incubated lactose broth pool in 100 ml of selective enrichments should be carried out. When it is important that the sample which contains *Salmonella* be recognized, then the samples are started in the usual way by incubation in lactose broth. After incubation, ten of the lactose broth samples may be pooled for incubation in the selective enrichment broth. This method uses the same amount of selective enrichment broth, but requires that fewer plates be streaked. The lactose broth which was not used in the pool from samples is refrigerated, and if positives are found, those broth samples included in the positive pool can be brought out of refrigeration and streaked to the selective medium.

It is essential that biochemical testing be carried out on pure cultures from selective culture medium plates. If mixed cultures are inoculated into the biochemical tube media, misleading results will be obtained, and it will be impossible to identify the culture to genus of *Salmonella* or to exclude it as a non-*Salmonella* organism. In many laboratories, it is deemed preferable to use one of the three approved commercial systems for presumptive generic identification of foodborne *Salmonella*. The three commercial systems which have been approved are API, Enterotube, and Minitek. When the commercial system is chosen, it should be checked in the laboratory where it is to be used to make sure that it will conform to the biochemical tube system which is available in that laboratory. The use of these commercial kits save much time and labor, and if correlation is found to be adequate, then the commercial system should prove to be preferable for tentative generic identification of isolates.

When cultures have been presumptively confirmed as *Salmonella*, then serological somatic (O) and flagellar (H) tests should be performed. When this is done, the cultures can then be classified into one of three groups: *Salmonella* when presumptive positive by commercial kits and when the culture is positive on the somatic (O) and (H) tests, not *Salmonella* when presumptively classified as not *Salmonella* with commercial biochemical test kit, or when cultures which do not conform to either of the above, then additional testing must be performed. In some laboratories not equipped to handle all the tests the cultures can be sent to a reference typing laboratory for definitive serotyping and identification. The latter step is often advisable simply to have additional evidence which can be used in critical epidemiological investigations.

B. *SALMONELLA* SEROLOGICAL TESTING FOR IDENTIFICATION

As discussed previously, *Salmonella* species and types are distinguished from one another on the basis of antigens which are present as a part of the structure of the cell walls and the flagella. These antigens are used to stimulate the production of antibodies in the blood serum of animals when injected into the animal. Most *Salmonella* antisera produced commercially is produced in the rabbit, because of less cost, and the fact that the rabbit produces a good response to the injection of these antigenic substances. Flagellar antigens in the *Salmonella* are designated as H antigens, cell wall or somatic antigens as O antigens, and the antigen comprising the envelope of some cells is designated the Vi antigen because it was originally thought that this might be the substance responsible for the virulence of the organism. When flagellar, somatic, and capsular antigens and antibodies are all present in a test preparation, the flagellar antigens and antibodies, as the external groups, react first, followed by reaction of the capsular, and then the somatic antigens. Because this type of reaction can be confusing, the antibodies are usually absorbed so that only one type of antigen-antibody will take place within a single test. In addition to removal of antigen types as above, absorption can also remove antibodies of different specificities so that only one flagellar or one somatic antibody will remain in an antiserum, and the specific identity of these materials can be established. For example, if a bacterium contains somatic (O) antigens A, B, D, and E, then any antiserum which contains antibodies a, b, d, or e will react with that cell, and it is not necessary that all antibodies be present to show a reaction. In this way, polyvalent antisera can be prepared so that unknown organisms can be grouped and fewer antisera need be used for identification. In such a situation, an antiserum (Poly A) can be obtained commercially which contains antibodies a, b, d, e, and l. Another (Poly C) can be obtained which contains antibodies i, j, k, m, n, and o. If one now tests an unknown strain isolated from a sample, and gets a reaction with Poly A serum, but not with Poly C, then one knows that the organism contains one of the antigens A, B, D, E, or L, but does not contain antigens I, J, K, M, N, or O. One then uses the monovalent antisera containing the corresponding antibodies only, for the identity of the somatic antigen present. Following determination of the specific somatic antigen present, the specific H antigen is determined in a similar manner. If, in the initial reaction, there is no reaction with any of the polyvalent O antisera, this does not mean that the organism is not *Salmonella*. Rather the O antigen may simply be covered by the presence of the Vi antigen, and an additional test must be carried out. In this case, the organism that does not react with the O antiserum is reacted with a Vi antiserum. If no reaction occurs, then it is concluded that the organism is not *Salmonella*.

Before application of serological techniques, biochemical identification of all isolates to genus level should have been completed. Serological identification of any bacterium, although very specific, is also comparatively very expensive and should therefore be a confirmatory procedure, not a screening procedure. For use in serological procedures, the cells used should not have been subjected to unusual stresses of any kind, particularly to heat stress. For serological testing, all materials and equipment should be at room temperature, and the dilution of the antiserum to be used should be verified as the proper dilution for use for that specific serum. Generally, serological testing for the *Salmonella* begins with the testing of the O antisera only, and H antisera is reserved for later use in specific identifications.

Polyvalent antisera are used initially in slide agglutination tests to group the organisms into specific antigenic combinations for the sake of preserving antiserum reagents. Each test run should include a saline, negative control, and a positive agglutination control in order that more accurate readings may be obtained. In rapid slide tests, living organisms are used as the test antigen, and this fact must always be remembered for disinfection of the equipment and the space used in the testing process. Following tests, all equipment used must be sterilized by autoclaving or by chemical treatment to avoid spread of the pathogenic organisms.

Commercial preparations of *Salmonella* O, H, and Vi antibodies are available.[29] These are provided as stable, freeze-dried antisera which are produced in rabbits. When required, the antisera have been absorbed to result in single factor or group sera for use. The antisera are provided as polyvalent group sera to allow reduction of the numbers of different antisera required for specific identification of an isolate. The naming of the polyvalent sera is sometimes confusing since letters of the alphabet are used to designate certain polyvalent groups, and these letters duplicate the *Salmonella* group designations given in the group designation scheme.

Although serological testing can be used to provide corroborative evidence of identity of cultures of *Salmonella* and other bacteria, these tests cannot be used alone to identify agents of disease, or bacteria isolated from contaminated food samples. Prior to this testing, biochemical and morphological evidence should have indicated that the organism in question was likely to be a member of the *Salmonella* genus, and the serological testing should be the last step in the procedure for the isolating laboratory. If any question remains, and to provide further corroborative evidence, the culture being tested should be sent to one of the FDA or USDA laboratories for identification. If similar results are obtained from two independent sources in this way, then it is likely that the identification is correct.

REFERENCES

1. **Speck, M. L., Ed.**, *Compendium of Methods For the Microbiological Examination of Foods*, American Public Health Association, Washington, D.C., 1976, 301.
2. **Moran, A. B., Ed.**, Isolation and identification of Salmonella from foods, in *Microbiology Laboratory Guidebook*, U.S. Department of Agriculture, Washington, D.C., 1974.
3. **Association of Official Analytical Chemists,***Official Methods of Analysis of the Association of Official Analytical Chemists*, 13th ed., Association of Official Analytical Chemists, Arlington, VA, 1980.
4. **Food and Drug Administration**, *Bacteriological Analytical Manual*, Association of Offical Analytical Chemists, Arlington, VA, 1984.
5. Selected Microbiological Methods, Center for Communicable Diseases, Atlanta, GA, 1968.
6. **D'Aoust, J.-Y. and Sewell, A. M.**, Detection of *Salmonella* with the Bio-enzabead™ enzyme immunoassay technique, *J. Food Prot.*, 51, 538, 1988a.
7. **D'Aoust, J.-Y. and Sewell, A. M.**, Reliability of the 1-2 test™ System for detection of *Salmonella* in foods, *J. Food Prot.*, 51, 853, 1988b.
8. **Eckner, K. F., Flowers, R. S., Robison, B. J., Mattingly, J. S., Gabis, D. A., and Silliker, J. H.**, Comparison of the Bio-enzabead™ immunoassay method and conventional culture procedure for detection of *Salmonella* in foods, *J. Food Prot.*, 50, 379, 1987.
9. **Fitts, R.**, Development of a DNA-DNA hybridization test for the presence of *Salmonella* in foods, *Food Technol.*, 39(3), 95, 1985.
10. **Flowers, R. S., Klatt, M. J., Mozola, M. A., Curiale, M. S.,. Gabis, D. A, and Silliker, J. H.**, DNA hybridization assay for detection of *Salmonella* in foods: collaborative study, *J. Assoc. Off. Anal. Chem.*, 70, 521, 1987a.
11. **Flowers, R. S., Klatt, M. J., Robison, B. J., Mattingly, J. A., Gabis, D. A., and Silliker, J. H.**, Enzyme immunoassay for detection of *Salmonella* in low moisture foods: collaborative study, *J. Assoc. Off. Anal. Chem.*, 70, 530, 1987b.
12. **Ibrahim, G. F. and Lyons, M. J.**, Detection of Salmonellae in foods with an enzyme immunometric assay, *J. Food Prot.*, 50, 59, 1987.
13. **Sall, B. S., Lombardo, M., Sheridan, B., and Parsons, G. H.**, Performance of a DNA probe-based *Salmonella* test in the AACC check sample program, *J. Food Prot.*, 51, 579, 1988.
14. **Committee on the Scientific Basis of the Nation's Meat and Poultry Inspection Program**, *Meat and Poultry Inspection. The Scientific Basis of the Nation's Program*, National Research Council, National Academy Press, Washington, D.C., 1985.
15. **Litchfield, J. H.**, Salmonella Food Poisoning, in *The Safety of Foods*, 2nd ed., Graham, H. D., Ed., AVI Publishing, Westport, CT, 1978, 120.
16. **National Research Council**, *Drinking Water and Health*, National Academy of Sciences, Washington, D.C., 1977.
17. **Graham, H. D.**, *Safety of Foods*, 2nd ed., AVI Publishing, Westport, CT, 1964.
18. **Troller, J. A.**, The water relations of food-borne bacterial pathogens. A review, *J. Milk Food Tech.*, 36, 276, 1973.
19. *Standard Methods for the Examination of Water and Wastewater*, 16th ed., American Public Health Association, Washington, D.C., 1985, 880.
20. **Nickerson, J. T. and Sinskey, A. J.**, *Microbiology of Foods and Food Processing*, American Elsevier, New York, 1972.
21. **Food and Drug Administration**, *Bacteriological Analytical Manual*, 6th ed., Association of Official Analytical Chemists, Arlington, VA, 1984.
22. **Guthrie, R. K. and Reeder, D. J.**, Membrane filter-fluorescent antibody method for the detection and enumeration of bacteria in water, *Appl. Microbiol.*, 17, 399, 1969.

23. **Abshire, R. and Guthrie, R. K.,** The use of fluorescent antibody techniques for detection of *Streptococcus faecalis* as an indicator of fecal pollution of water, *Water Res.*, 5(11), 1089, 1971.

24. **Abshire, R. and Guthrie, R. K.,** Fluorescent antibody as a method for the detection of fecal pollution: *Escherichia coli* as indicator organisms, *Can. J. Microbiol.*, 19(2), 201, 1973.

25. **Association of Official Analytical Chemists,** AOAC-Approved Method Screens Salmonella Samples in 48 Hours, in *Prepared Foods*, Association of Official Analytical Chemists, Arlington, VA, 1988.

27. **Nath, E. J., Neidert, E., and Randall, C. J.,** Evaluation of enrichment protocols for the 1-2 Test™ for Salmonella detection in naturally contaminated foods and feeds, *J. Food Prot.*, 52(7), 498, 1989.

27. **Gene-Trak Systems,** *Gene-Trak Salmonella Assay*, Gene-Trak Systems, Framingham, MA, 1988.

28. **Difco Laboratories,** *Difco Manual*, Difco Laboratories, Detroit, 1984.

Appendix A

CULTURE MEDIA USED FOR *SALMONELLA* ISOLATION, CULTURE, AND IDENTIFICATION

The need for culture media of different types depends upon the foods being tested, the amount of contamination of all types expected, and the stage of testing at which the medium is needed. The kinds of culture media used vary from the general purpose culture medium to the highly specialized, selective-differential media used for identifications, and include specially formulated media for detection of fermentation and biochemical utilization of specific metabolic requirements of some bacteria.

A general purpose culture medium is one which will support the growth of a number of different kinds of bacteria under most ordinary culture conditions. Such a medium contains more nutrient sources than are required for most of the organisms which will be encountered in environments likely to be tested. Such a medium is included in this listing as Brain Heart Infusion (broth or agar).

A selective culture medium supports the growth of certain bacteria while inhibiting the growth of others, particularly those that may be present in large numbers that are likely to be found in the same environment. Selective media are frequently used to enrich for certain desired bacteria when testing food or other samples for the bacteria present. While many selective media were developed for use in the clinical laboratory, many others have been developed for use in the culture of foods of a certain nature, because the ingredients of the medium will enhance the growth of the desired organisms.

The methods now best used for isolation and identification of *Salmonella* from foods are included in five basic steps. These include (1) the initial step of pre-enrichment which allows the organisms to be exposed to a nutritious medium for a period of time to recover from any stress which may have been applied up to this point, and permits injured cells to recover. (2) The selective enrichment step in which the sample is allowed exposure to a culture medium which favors growth of the desired bacterial culture (*Salmonella*). This medium will contain some ingredient which will not only promote the growth of these bacteria, but will selectively inhibit those bacteria which are expected to be present in the same environment. Selective enrichment allows the growth of *Salmonella* to continue from the pre-enrichment while other bacterial growth is inhibited. (3) From this step, when the sample is plated to solid selective media, there are fewer extraneous bacteria present, and it is much easier to detect and identify colonies suspected to be *Salmonella*. (4) From the selective solid media, suspect colonies can be readily picked to biochemical screening media which further identifies the isolates to provide at least tentative generic identification of *Salmonella* cultures. (5) The fifth step in identification consists of use of serological techniques for specific identification of the cultures.

The *Bacteriological Analytical Manual*[3] was supplemented in September 1987 with new methods for rapid detection of *Salmonella* in foods. While these methods are useful for screening certain foods — for example, the hydrophobic grid-membrane filter method is applicable to detection of the organisms in chocolate, raw poultry meat, pepper, cheese powder, powdered egg, and instant nonfat dry milk — and are more rapid than the complete, traditional, culture procedures followed by serological procedures, they are still useful only as screening methods, and identifications still must be finalized with standard culture, serological, and/or biochemical methods as specified by the method used. All of the newer, more rapid methods still require the use of at least some of the same reagents and culture media. For the reagents and/or culture media needed, consult the supplement to the 6th edtion of *Bacteriological Analytical Manual*.[3]

For the enzyme immunoassay method of detection of *Salmonella* in foods, the antibody preparation used must be specific for the antigen expected to avoid false positive reactions. The DNA probe method allows the identification of *Salmonella*-free samples within 48 h, but positive assays must be confirmed by the traditional cultural methods and serological techniques.

In most laboratories at the present, media are available in dehydrated form to be prepared as needed in the laboratory. Most of those media will be identical to the ones listed here, although in some there may be slight modifications which have been found to improve culture growth as the media were used over the years. In some instances, the commercial houses have found that slight changes in pH, electrolyte concentration, or ingredient amounts improve the performance of the medium, and these minor changes have been incorporated for general use. Particularly in regulatory work, or in maintenance of quality control within a plant, it is advisable to inoculate known controls when medium lots are being changed. Such controls are not often used in general culture work, but would improve specificity and laboratory performance if included more often.

Several general points should be fully realized by all laboratory personnel in respect to culture medium usage. Among these points, perhaps the most important is to use a culture medium source, for either dehydrated or fresh media which is reliable. Less expensive products may be less effective products, and only those recognized medium sources should be used. More problems may be encountered when using selective media than with others because of the content of ingredients which may have toxic effects on some organisms. In these media, for example, only slight overheating during preparation may cause excessive toxicity, even to the organisms which normally grow on the medium. It is important to realize that not all bacteria will grow equally well on all ingredients. For this reason, some fermentation media have different formulations of the broth base, or the indicator system used. In some organisms, it is also important to remember that not all media are suitable for the production of some toxins or antigens by the organisms being studied.

A. MEDIA FORMULAS

Bismuth Sulfite (BS) agar (Wilson and Blair)

Peptone	10 g
Beef extract	5 g
Dextrose	5 g
Na_2HPO_4 (anhydrous)	4 g
$FeSO_4$ (anhydrous)	0.3 g
Bismuth sulfite (indicator)	8 g
Brilliant Green	0.025 g
Agar	20 g
Distilled water	1 liter

Mix thoroughly and heat while stirring. Boil about 1 min to obtain uniform suspension. (Precipitate will not dissolve.) Cool to 45 to 50°C. Suspend precipitate by gentle mixing and pour into sterile petri dishes. Let plates dry with lids propped open until obvious liquid is removed from the surface of the agar. Final pH should be 7.6. *DO NOT AUTOCLAVE.* Media should be prepared 1 day before use, but should not be used more than 2 days after preparation because of decreased selectivity.

Brain Heart Infusion (BHI) Agar and Broth

Calf brain infusion	200 g
Beef heart infusion	250 g
Proteose peptone	10 g
NaCl	5 g
$Na_2HPO_4 \cdot H_2O$	2.5 g
Dextrose	2 g
Distilled Water	1 liter

These ingredients are dissolved and dispensed as needed for culture in flasks or tubes, then autoclaved for 15 min at 121°C. pH should be 7.4.

To prepare Brain Heart Infusion agar, add 15 g granular agar to the above and melt by boiling gently. Autoclave for 15 min at 121°C to sterilize. Pour into plates when the agar has cooled sufficiently. Leave plate lids propped open for a short time to allow excess moisture to evaporate from plate surfaces. Store plates inverted after setting.

Brilliant Green Agar (BGA)

Peptone	10 g
Yeast extract	3 g
Sodium chloride	5 g
Lactose	10 g
Sucrose	10 g
Phenol red; 0.2 % aqueous solution	40 ml
Brilliant green, 1.0% aqueous solution	12.5 ml
Agar	15 g
Distilled water	1 liter

Autoclave at 121°C for 15 min. *AVOID OVERHEATING*. Final pH should be 6.9. The medium will be orange in color. Pour into relatively thick plates and dry by leaving lids propped open for a time. Overheating or reheating will result in loss of selectivity.

Gelatinase solution, 5%

Gelatinase	5 g
Distilled water	100 g

Suspend gelatinase in distilled water, then centrifuge for 10 min at 9500 rpm. Filter the supernatant through a 0.45 µm membrane filter. Dispense into screw-capped bottles, aseptically.

Hektoen Enteric (HE) agar

Proteose peptone	12 g
Yeast extract	3 g
Bile salts	9 g
Lactose	12 g
Sucrose	12 g
Salicin	2 g
NaCl	5 g
Sodium thiosulfate	5 g
Ferric ammonium citrate	1.5 g
Bromthymol blue	0.064 g
Acid fuchsin	0.1 g
Agar	13.5 g
Distilled water	1 liter

Heat to boiling with stirring. Do not boil longer than 1 min. Cool and pour into petri dishes for use. Let dry with lids propped open until excess liquid is removed from surface of plates. Final pH should be 7.6. Use the day of preparation.

Indole medium

Indole production is detected by use of 2% tryptone solution or by use of trypticase (tryptic) soy broth. Both media contain sufficient tryptophane for determination of indole production. These media are listed below. For trypticase soy broth see page 167.

Kligler Iron Agar

Beef extract	3 g
Yeast extract	3 g
Peptone	15 g
Proteose peptone	5 g
(Polypeptone peptone in the amount of 20 g	
may be substituted for the above 4 ingredients.)	
Lactose	20 g
Dextrose	1 g
NaCl	5 g
Ferric ammonium citrate	0.5 g
Sodium thiosulfate	0.5 g
Agar	15 g
Phenol red	0.025 g
Distilled water	1 liter

Heat with stirring to dissolve. Dispense into 13 × 100 mm screw-capped tubes to autoclave for 15 min at 121°C. Cool in a slanted position to form deep butts of medium. pH should be 7.4.

Kligler Iron agar may be used instead of Triple Sugar Iron agar as the user prefers. Triple Sugar Iron agar contains essentially the same ingredients plus sucrose, and therefore differentiates those organisms which ferment that sugar but not lactose. Color of reaction may vary somewhat between the two media, but both will be of use in distinguishing *Salmonella* from other enteric bacteria.

Lactose Broth

Beef extract	3 g
Peptone	5 g
Lactose	5 g
Distilled water	1 liter

Dissolve ingredients and dispense into tubes with fermentation tubes or into flasks in 225 ml amounts as required for use. Autoclave for 15 min at 121°C. Final pH should be 6.9.

Lauryl Tryptose (LST) Broth

Tryptose (trypticase)	20 g
Lactose	5 g
K_2HPO_4	2.75 g
KH_2PO_4	2.75 g
NaCl	5 g
Sodium lauryl sulfate	0.1 g
Distilled water	1 liter

Dispense in screw-capped tubes with fermentation tubes and autoclave for 15 min at 121°C. Final pH should be 6.8.

Lysine Decarboxylase broth (Falkow) (for *Salmonella*)

Peptone	5 g
Yeast extract	3 g
Glucose	1 g
L-lysine	5 g
Bromcresol purple	0.02 g
Distilled water	1 liter

Heat to dissolve. Dispense in 5 ml portions in screw-capped tubes. Autoclave for 15 min at 121°C. Final pH should be between 6.5 and 6.8.

Lysine Iron agar (Edwards and Fife)

Peptone	5 g
Yeast extract	3 g
Dextrose	1 g
L-lysine	10 g
Ferric ammonium citrate	0.5 g
Sodium thiosulfate (anhydrous)	0.04 g
Bromcresol purple	0.02 g
Agar	15 g
Distilled water	1 liter

Heat to dissolve ingredients. Dispense in 4 ml amounts in 13×100 screw-capped tubes. Autoclave for 12 min at 121°C. Slant for cooling to form butts of 4 cm and slants of 2.5 cm. Final pH should be 6.7.

MR — VP medium (Voges-Proskauer medium)

Peptone	7 g
NaCl	5 g
Dextrose	5 g
Distilled water	1 liter

Dissolve ingredients and if necessary adjust pH to 6.5. Autoclave for 10 min at 121°C after dispensing to screw-capped tubes in 5-ml amounts.

MacConkey (MAC) agar

Proteose peptone	3 g
Peptone	17 g
Lactose	10 g
Bile salts No 3	1.5 g
NaCl	5 g
Neutral red	0.03 g
Crystal violet	0.001 g
Agar	13.5 g
Distilled water	1 liter

Heat with stirring to dissolve. Autoclave 15 min at 121°C. Cool and pour into petri dishes for use. Leave lids propped open to allow excess moisture to evaporate from surface of plates. Final pH should be 7.1.

Malonate broth

Yeast extract	1 g
$(NH_4)_2SO_4$	2 g
K_2HPO_4	0.6 g
KH_2PO_4	0.6 g
NaCl	2 g
Sodium malonate	3 g
Dextrose	0.25 g
Bromthymol blue	0.025 g
Distilled water	1 liter

Dissolve. Heat if necessary. Dispense 3 ml into 13 × 100 screw-capped tubes. Autoclave for 15 min at 121°C. Final pH should be 6.7.

Motility test medium (semisolid)

Beef extract	3 g
Peptone	10 g
NaCl	5 g
Agar	4 g
Distilled water	1 liter

Heat with stirring to dissolve agar. Autoclave for 15 min at 121°C to sterilize after dispensing in 5 to 10 ml amounts in screw-capped test tubes. Final pH should be 7.4.

Nonfat dry milk (reconstituted)

Nonfat dry milk	100 g
Distilled water	1 liter

For *Salmonella*, dissolve and dispense 225 ml portions into flasks. Autoclave for 15 min at 121°C. Just before use, aseptically readjust volume to 225 ml with sterile distilled water.

Nutrient broth

Beef extract	3 g
Peptone	5 g
Distilled water	1 liter

Heat to dissolve. Dispense into 10 ml portions in tubes, or 225 ml portions in flasks as needed for cultures. Autoclave for 15 min at 121°C. Final pH should be 6.8.

Phenol Red Carbohydrate broth

Proteose peptone No 3	10 g
NaCl	5 g
Beef extract	1 g
Phenol red	0.018 g
Distilled water	1 liter

Dissolve 5 g of dulcitol, 2 g of lactose, or 10 g of sucrose as required for *Salmonella* test in the broth. Dispense into small tubes with inverted fermentation tubes and autoclave for 10 min at 110°C. (Normally 12 lb pressure — test your autoclave for proper function. Do not overheat.) If unable to control heat, sterilize medium by filtration to assure proper reaction of carbohydrates.

Potassium Cyanide (KCN) broth

Potassium cyanide	0.5 g
Proteose peptone No. 3	3 g
NaCl	5 g
KH_2PO_4	0.225 g
Na_2HPO_4	5.64 g
Distilled water	1 liter

Dissolve all ingredients except KCN and autoclave for 15 min at 121°C. Cool to 5 to 8°C. Dissolve 0.5 g KCN stock solution in 100 ml sterile distilled water cooled to 5 to 8°C. *USE BULB PIPETTER ONLY — DO NOT MOUTH PIPETTE* add 15 ml of cold KCN solution to 1 liter sterile cold base. *DO NOT PIPETTE BY MOUTH.* Mix and aseptically transfer 1 to 1.5 ml portions to

13×100 tubes. Aseptically stopper with No. 2 corks impregnated with paraffin. Do not allow paraffin to flow into broth, but it should form a seal between the cork and tube. Store tubes at 5 to 8°C for no longer than 2 weeks before use.

Purple Carbohydrate broth base

Proteose peptone No. 3	10 g
Beef extract	1 g
NaCl	5 g
Bromcresol purple	0.02 g
Distilled water	1 liter

This broth base is prepared as is phenol red carbohydrate broth base, and carbohydrates are added as needed in the same manner. As in that case, it is important that the medium not be overheated, and if necessary, final mixtures should be filtered to sterilize.

Selenite Cystine Broth

This medium may be made by either of the following formulae. The user is urged to use that formulation which is best suited to individual use. If directions are followed precisely, results should be consistent.

Formulation 1

Tryptone	5 g
Lactose	4 g
Sodium selenite ($NaHSeO_3$)	4 g
Na_2HPO_4	10 g
L-cystine	0.01 g
Distilled water	1 liter

Heat to boiling to dissolve. Dispense in 10 ml amounts into sterile screw-capped tubes. If needed or desired, may be heated for 10 min in flowing steam. *DO NOT AUTOCLAVE*. Final pH should be 7.0. Medium is not sterile and should be used the same day as prepared.

Formulation 2 (North-Bartram modification)

Polypeptone	5 g
Lactose	4 g
Sodium selenite ($NaHSeO_3$)	4 g
Na_2HPO_4	5.5 g
KH_2PO_4	4.5 g
L-cystine	0.01 g
Distilled water	1 liter

Heat to boiling to dissolve. Dispense in 10 ml amounts into sterile screw-capped tubes. If needed or desired, may be heated for 10 min in flowing steam. *DO NOT AUTOCLAVE.* Medium is not sterile and should be used the same day as prepared.

Simmons Citrate agar

Sodium citrate.$2H_2O$	2 g
NaCl	5 g
K_2HPO_4	1 g
$NH_4H_2PO_4$	1 g
$MgSO_4$	0.2 g
Bromthymol blue	0.08 g
Agar	15 g
Distilled water	1 liter

Heat to boiling to melt agar. Dispense into screw-capped tubes in approximately 6 to 8 ml amounts. Autoclave for 15 min at 121°C. Slant tubes for cooling to solidify. Final pH should be 6.9.

Tetrathionate broth

Proteose peptone	5 g
Bile salts	1 g
Calcium carbonate	10 g
Sodium thiosulfate.$5H_2O$	30 g
Distilled water	1 liter

Heat to boiling and mix. Precipitate will not completely dissolve. Add 20 ml iodine solution (6 g iodine crystal plus 5 g potassium iodide in 20 ml distilled water). On day of use, aseptically add 10 ml of brilliant green solution prepared by solution of 0.1 g sterile brilliant green dye to 100 ml sterile distilled water. Do not heat medium after adding iodine and brilliant green. Dispense in 10 ml amounts to sterile screw-capped tubes. Use medium same day prepared.

Thioglycollate medium

Trypticase or tryptone	15 g
Phytate peptone	3 g
Dextrose	6 g
NaCl	2.5 g
Sodium thioglycollate	0.5 g
L-cystine	0.25 g
Na_2SO_4	0.1 g
Agar	0.7 g
Distilled water	1 liter

Heat with stirring to dissolve agar. Fill screw-capped tubes one half full. Autoclave for 15 min at 118°C. Final pH should be 7.0.

Triple Sugar Iron (TSI) agar

Beef extract	3 g
Yeast extract	3 g
Peptone	15 g
Proteose peptone	5 g
Glucose	1 g
Lactose	10 g
Sucrose	10 g
$FeSO_4$	0.2 g
NaCl	5 g
$Na_2S_2O_3$	0.3 g
Phenol red	0.024 g
Agar	12 g
Distilled water	1 liter

Heat to boiling to dissolve ingredients. Fill screw-capped tubes approximately one third full. Autoclave for 15 min at 121°C. Before medium cools, slant tubes to obtain a butt of approximately 3 cm. Final pH should be 7.4.

Trypticase soy — Tryptose broth

Trypticase soy broth (commercial, dehydrated)	15 g
Tryptose broth (commercial, dehydrated)	13.5 g
Yeast extract	3 g
Distilled water	1 liter

Heat gently to dissolve ingredients. Dispense in screw-capped tubes in 5 ml amounts. Autoclave for 15 min at 121°C. Final pH should be 7.2.

Tryptone

Granular tryptone	15 g
Distilled water	1 liter

Autoclave for 15 min at 121°C.

Urea broth

Urea	20 g
Yeast extract	0.1 g
KH_2PO_4	9.1 g
Na_2HPO_4	9.5 g
Phenol red	0.01 g
Distilled water	1 liter

Dissolve. Do not heat. Sterilize by filtration through 0.45 μm membrane. Aseptically dispense 1.5 to 3.0 ml into sterile screw-capped test tubes. Final pH should be 6.8.

Xylose Lysine Desoxycholate (XLD) agar

Yeast extract	3 g
L-lysine	5 g
Xylose	3.75 g
Lactose	7.5 g
Sucrose	7.5 g
Sodium desoxycholate	2.5 g
Ferric ammonium citrate	0.8 g
Sodium thiosulfate	6.8 g
NaCl	5 g
Agar	15 g
Phenol red	0.08 g
Distilled water	1 liter

Heat with stirring to boiling. Do not overheat. Pour into plates and let dry with lids propped open for about 2 h, or until excess moisture has evaporated from agar. Final pH should be 7.4. Do not store for more than 1 day.

REFERENCES FOR OTHER MEDIA AND METHODS

1. **Bailey, W. R. and Scott, E. G.,** *Diagnostic Microbiology*, 4th ed., C. V. Mosby, St. Louis, 1974.
2. **Difco Laboratories,** *Difco Manual*, 10th ed., The Difco Company, Detroit, 1984.
3. **Association of Official Analytical Chemists,** *Bacteriological Analytical Manual*, Food and Drug Administration, Association of Official Analytical Chemists, Arlington, VA, 1984.
4. **Harrigan, W. F. and McCance, M. E.,** *Laboratory Methods in Food and Dairy Microbiology*, Revised ed., Academic Press, New York, 1976.

Appendix B

ANTIGENIC SCHEMA FOR *SALMONELLA* (KAUFFMAN-WHITE SCHEMA, MODIFIED)

The following Schema was taken from *Difco Manual*.[1] It is the modification of the original Schema as proposed by Ewing.[2]

ALPHABETICAL LIST OF THE SEROTYPES OF *SALMONELLA*

Organism	O Group
Salmonella enteritidis	
ser Aba 6,8:i:e,n,z	C2
ser Abadina 8:g,m:[e,n,z15]	M
ser Abaetetuba 11:k:1,5	F
ser Aberdeen 11:i:1,2	F
ser Abony 1,4,5,12:b:e,n,x	B
variant Haifa 4,12:b:e,n,x	B
ser Abortusbovis 1.4.12.27"b"e.m.x	B
ser Abortuscanis 4,5,12:b:z5	B
ser Abortussequi 4,12:-:e,n,x	B
ser Abortusovis 4,12:c:1,6	B
ser Accra 1,3,19:b:z6	E4
*ser Acres 1,13,23:b:z42:[1,5]	G2
ser Adabraka 3,10:z4,z23:[1,7]	E1
ser Adamstown 28:k:1,6	M
ser Adamstua 11:e,h.1,6	F
ser Adalaide 35:f,g:-	O
ser Adeoyo 16:g,m:-	I
ser Aderike 28:z38:-	M
ser Adjame 13,23:r:1,6	G2
ser Aequatoria 6,7:z4,z23:e,n,z15	C1
ser Aertrycke = ser Typhimurium	
ser Aflao 1,6,14,25:l,z28:e,n,x	H
ser Africana 4,12:i:1,6	B
ser Agama 4,12:i:1,6	B
ser Agbeni 13,23:g,m:-	G2
ser Agege 3,10:c:e,n,z15	E1
ser Ago 30:z38:-	N
ser Agodi 35:g,t:-	O
ser Agona 4,12;f,g,s:-	B
ser Ahmadi 1,3,19:d:1,5	E4
ser Ahuza 43:k:1,5	U
ser Ajiobo 13,23:z4,z23:-	G2
ser Akanji 6,8:r:1,7	C2
ser Akuafo 16:y:1,6	
ser Alabama 9,12:c:e,n,z15	D1
ser Alachua 35:z4,z23	O

Organism	O Group
ser Alagbon 6,8:y:1,7	C2
ser Alamo 6,7:g,z51:1,5	C1
ser Albany (8),20:z4,z24:-	C3
ser Albert 4,12:z10:e,n,x	B
ser Albuquerque 6,14,24:d:z6	H
*ser Alexander 3,10:z:1,5	E1
ser Alexanderplatz 47:z38:-	X
ser Alexanderpolder (8):c:l,w	C3
ser Alger 38:l,v:1,2	P
ser Allandale 1,40:k:1,6	R
ser Allerton 3,10:b:1,6	E1
*ser Alsterdorf ,140:g,m,t:-	R
ser Altendorf 4,12:c:1,7	B
ser Altona (8),20:c:1,7	C3
ser Amager 3,10:y:1,2	E1
ser Amba 11:k:l,z13,z28	F
ser Amersfoort 6,7:d:e,n,x	C1
ser Amherstiana (8):l,(v):1,6	C3
ser Amina 16:i:1,5	I
ser Aminatu 3,10:a:1,2	E1
ser Amounderness 3,10:i:1,5	E1
ser Amoutive 28:d:1,5	M
ser Amsterdam 3,10:g,m,s:-	E1
ser Amunigun 16:a:1,6	I
ser Anatum 3,10:e,h:1,6	E1
ser Anderlecht 3,10:c:l,w	E1
ser Anfo 39:y:1,2	Q
ser Angoda 30:k:e,n,x	N
*ser Angola 1,9,12:z:z6	D1
ser Ank 28:k:e,n,z15	M
ser Annedal 16:r,(i):e,n,x	I
ser Antonio 57:a:z6	57
ser Antsalova 51:z:1,5	51
ser Apapa 45:m,t:-	W
ser Aqua 30:k:1,6	N
ser Ardwick 6,(7),(14):f,g:-	C1
ser Arechavaleta 4,5,12:a:[1,7]	B
*ser Argentina 6,7:z36:-	C1
ser Arkansas (3),(15),34:e,h:1,5	E3
*ser Artis 56:b:-	56
ser Aschersleben 30:b:1,5	E3
ser Ashanti 28:b:1,6	M
*ser Askraal 51:l,z28:-	51
ser Assen 21:a:-	L
ser Atherton = ser Waycross	
ser Atlanta (Mississippi)[1],13,23:b:-	G2
*ser Atra 50:m,t:z6:z42	Z
ser Augustenborg 6,7:i:1,2	C1
ser Austin 6,7:a:1,7	C1
ser Avonmouth 1,3,19:i:e,n,z15	E4

Organism	O Group
ser Ayinde 4,12,27:d:z6	B
ser Ayton 1,4,12,27:l,w:z6	B
ser Azteca 4,5,12:l,v:1,5	B
ser Babelsberg 28:z4,z23:e,n,z15	M
*ser Bacongo 6,7:z36:z42	C1
ser Baguirmi 30:y:e,n,x	N
ser Bahati 13,22:b:e,n,z15	G1
ser Bahrenfeld 6,14,24:e,h:1,5	H
ser Baiboukoum 6,7:k:1,7	C1
ser Baildon (9),46:a:e,n,x	D2
ser Ball 1,4,12,27:y:e,n,x	B
ser Bambesa, Joined w/ser Miami	
ser Bamboye (9),46:b:l,w	D2
ser Banalia 6,8:b:z6	C2
ser Banana, Joined w/ser California	
ser Bandia 35:i:l,w	O
ser Bantam = ser Meleagridis	
*ser Baragwanath 6,8:m,t:1,5	C2
ser Bardo (8):e,h:1,2	C3
ser Bareilly 6,7,[14]:y:1,5	C1
ser Barmbek 16:d:z6	I
ser Barranguilla 16:d:z6	I
*ser Basel 58:l,z13,z28:1,5	58
ser Batavia = ser Lexington	
*ser Bechuana 4,12,27:g,t:-	B
ser Bedford 1,3,19:l,z13,z28:e,n,z15	E4
ser Belem 6,8:c:e,n,x	C2
*ser Bellville 16:e,n,x:1,7	I
*ser Beloha 18:z36:-	K
ser Benfica 3,10:b:e,n,x	E1
variant T1 T1:b:e,n,x	
ser Benguella 40:b:z6	R
ser Bere 47:z4,z23:z6	X
ser Bergedorf (9),46:e,h:1,2	D2
ser Bergen 47:i:e,n,z15	X
ser Berkeley 43:a:1,5	U
ser Berlin 17:d:1,5	J
*ser Bern 1,40:z4,z32:-	R
ser Berta 9,12:f,g,t:-	D1
*ser Betioky 59:k:(z)	59
ser Biafra 3,10:z10:z6	E1
*ser Bilthoven 47:a:[1,5]	X
ser Bilu (1),3,10,(19):f,g,t:1,(2),7	E4
ser Binza 1,15:y:1,5	E2
ser Birkenhead 6,7:c:1,6	C1
ser Birmingham 3,10:d:l,w	E1
ser Bispebjerg 1,4,5,12:a:e,n,x	B
*ser Blankenese 1,9,12:b:z6	D1
*ser Bleadon 17:(f),g,t:[e,n,x,z15]	J
ser Bledgam 9,12:g,m,q:-	D1

Organism	O Group
ser Blijdorp 1,6,14,25:c:1,5	H
ser Blockley 6,8:k:1,5	C2
*ser Bloemfontein 6,7:b:[e,n,x]:z42	C1
ser Blukwa 18:z4,z24:-	K
ser Bobo 44:d:1,5	V
ser Bochum 4,5,12:r:l,w	B
*ser Bokenheim 1,53:z36,z38:-	53
ser Bodjonegoro 30:z4,z24:-	N
ser Boecker [1],6,14,[25]:l,v:1,7	H
ser Bokanjac 28:b:1,7	M
*ser Boksburg 40:g,s:e,n,x,z15	R
ser Bolombo 3,10:z38:-	E1
ser Bolton 3,10:y:e,n,z15	E1
ser Bombay, not confirmed	
*ser Bonaire 50:z4,z32:-	Z
ser Bonames 17:a:1,2	J
ser Bonariensis 6,8:i:e,n,x	C2
ser Bongor 48:z35:-	Y
ser Bonn 6,7:l,v:e,n,x	C1
ser Bootle 47:k:1,5	X
ser Borbeck 13,22:l,v:1,6	G1
*ser Bornheim 1,6,14,25:z10:1,(2),7	H
ser Bornum 6,(7),(14):z38:-	C1
*ser Boulders 13,23:m,t:z42	G2
ser Bournemouth 9,12:e,h:1,2	D1
ser Bousso 1,6,14,25:z4,z23:-	H
ser Bovismorbificans 6,8:r:1,5	C2
ser Bracknell 13,23,b:1,6	G2
ser Bradford 4,12,27:r:1,5	B
ser Braenderup6,7:e,h:e,n,z15	C1
ser Brancaster 1,4,12,27:z29:-	B
ser Brandenburg 4,12:l,v:e,n,z15	B
ser Brazil 16:a:1,5	I
ser Brazzaville 6,7:b:1,2	C1
ser Bredeney 1,4,12,27:l,v:1,7	B
*ser Bremen 45:g,m,s,t:e,n,x	W
ser Breukelen 6,8:l,z13:e,n,z15	C2
ser Brijbhumi 11:i:1,5	F
ser Brisbane 28:z:e,n,z15	M
ser Bristol 13,22:z:1,7	G1
ser Bron 13,22:g,m:[e,n,z15]	G1
ser Bronx 6,8:c:1,6	C2
ser Broughton 1,3,19:b:l,w	E4
ser Broxbourne = ser Wien	
ser Brunei (8),20:y:1,5	C3
ser Budapest 1,4,12:g,t:-	B
ser Buenosaires = ser Bonairiensis	
ser Bukavu 1,40:l,z28:1,5	R
ser Bukuru 6,8:b:l,w	C2
*ser Bulawayo 1,40:z:1,5	R

Organism	O Group
ser Bulbay 11:l,v:e,n,z15	F
*ser Bunnik 43:z42:[1,5]	U
ser Burgas 16:l,v:e,n,z15	I
ser Bury 4,12,27:c:z6	B
ser Businga 6,7:z:e,n,z15	C1
ser Butantan 3,10:b:1,5	E1
ser Buzu 1,6,14,25:i:1,7	H
ser Cairina 3,10:z35:z6	E1
ser Cairns 45:k:e,n,z15	W
ser Cairo 1,4,12,27:d:1,2	B
ser Calabar 1,3,19:e,h:l,w	E4
*ser Caledon 4,12:g,m:e,n,x	B
ser California 4,5,12:m,t:-	B
*ser Calvinia 6,7:a:z42	C1
ser Camberene 35:z10:1,5	O
ser Cambridge 3,15:e,h:l,w	E2
ser Canada 4,12:b:1,6	B
*ser Canastel 9,12:z29:1,5	D1
ser Canoga (3),(15),34:g,s,t:-	E3
ser Cannstatt 1,3,19:m,t:-	E4
*ser Cape 6,7:z6:1,7	C1
ser Caracas 1,6,14,25:g,m,s:-	H
ser Cardiff 6,7:k:1,10	C1
*ser Carletonville 38:d:[1,5]	P
ser Carmel 17:l,v:e,n,x	J
ser Carno 1,3,19:z:l,w	E4
ser Carrau 6,14,24:y:1,7	H
ser Casablanca 45:k:1,7	W
*ser Ceres 28:z;z39	M
ser Cerro 18:z4,z23:[z45]	K
ser Ceyco (9),46:k:z35	D2
ser Chagoua 1,13,23:a:1,5	G2
ser Chailey 6,8:z4,z23:[e,n,z15]	C2
*ser Chameleom 16:z4,z32:-	I
ser Champaign 39:k:1,5	Q
ser Chandans 11:d:e,n,x	F
ser Charity 1,6,14,25:d:e,n,x	H
*ser Chersina 47:z:z6	X
ser Chester 4,5,12:e,h:e,n,x	B
ser Chiba, not a Salmonella	
ser Chicago 28:r:1,5	M
ser Chincol 6,8:g,m,s:e,n,x	C2
ser Chingola 11:e,h:1,2	F
*ser Chinovum 42:b:1,5	T
ser Chittagong (1),3,10,(19):b:z35	E4
Salmonella cholerae-suis 6,7:c:1,5	C1
bioser Kunzendorf 6,7:[c]:1,5	C1
S. enteritidis	
ser Christiansborg 44:z4,z24:-	V
* variant 44:z4,z24:-s	V

Organism	O Group
*ser Chudleigh 3,10:e,n,x:1,7	E1
ser Clackamas 4,12:l,v,(x13):1,6	B
ser Claibornei 1,9,12:k:1,5	D1
ser Clerkenwell 3,10:e,n,x:1,7	E1
ser Cleveland 6,8:z10:1,7	C2
*ser Clifton 13,22:z29:1,5	G1
*ser Clovelly 1,44:z39:[e,n,x,z15]	V
ser Cocody (8),20:r,(i):e,n,z15	C3
ser Coeln 4,5,12:y:1,2	B
ser Coleypark 6,7:a:l,w	C1
ser Colindale 6,7:r:1,7	C1
ser Colombo 38:y:1,6	P
ser Colorado 6,7:w:1,5	C1
ser Concord 6,7:l,v:1,2	C1
ser Congo 13,23:g,t:-	G2
*ser Constantia 17:z:l,w:z42	J
ser Cook 39:z48:1,5	Q
ser Coquilhatville 3,10:z10:1,7	E1
ser Corvallis (8),20:z4,z23:-	C3
ser Cotham 28:i:1,5	M
ser Croft 28:g,m,s:-	M
ser Cuba = ser cubana	
ser Cubana 1,13,23:z29:-	G2
ser Curacao 6,8:a:1,6	C2
ser Dahlem 48:k:e,n,z15	Y
ser Dakar 28:a:a,6	M
ser Dalat, joined w/ser Ball	
ser Dallgow 1,3,19:z10:e,n,z15	E4
ser Dan 51:k:e,n,z15	51
*ser Daressalaam 1,9,12:l,w:e,n,x	D1
ser Daytona 6,7:c:1,5	C1
bioser Decatur 6,7:c:1,5	C1
*ser Degania 40:z4,z24:-	R
ser Dembe 35:d:l,w	O
ser Demerara 13,23:z10:l,w	G2
ser Denver 6,7:a:e,n,z15	C1
ser Derby 1,4,5,12:f,g:[1,2]	B
ser Dessau (1),3,15,(19):g,s,t:-	E4
*ser Detroit 42:z:1,5	T
ser Deversoir 45:c:e,n,x	W
ser Diguel 1,13,22:d:e,n,z15	G1
ser Diourbel 21:i:1,2	L
ser Djakarta 48:z4,z24:-	Y
ser Djermaia 28:z29:-	M
ser Djugu 6,7:z10:e,n,x	C1
ser Doncaster 6,8:a:1,5	C2
ser Donna 30:l,v:1,5	N
ser Dougi 50:y:1,6	Z
ser Dresden 28:c:e,n,x	M
ser Driffield 1,40:d:1,5	R

Organism	O Group
ser Drypool 3,15:g,m,s:-	E2
ser Dublin 1,9,12:g,p:-	D1
variant Vi+ 1,9,12,Vi:g,p:-	D1
*ser Dubrovnik 41:z:1,5	S
ser Duesseldorf 6,8:z4,z24:-	C2
ser Dugbe 45:d:1,6	W
ser Duisburg [1],4,12,[27]:d:e,n,z15	B
ser Durban 9,12:a:e,n,z15	D1
*ser Durbanville [1],4,12,[27]:[z39]:1,5,7	B
ser Durham 13,23:b:e,n,z15	G2
ser Duval 1,40:b:e,n,z15	R
ser Ealing 35:g,m,s:-	O
ser Eastbourne 1,9,12:e,h:1,5	D1
ser Eberswalde 28:c:1,6	M
ser Ebrie 35:g,m,t:-	O
ser Echa 38:k:1,2	P
ser Edinburg 6,7:b:1,5	C1
ser Edmonton 6,8:l,v:e,n,z15	C2
ser Egusi 41:d:-	S
ser Egisotpp 1,42:b:z6	T
*ser Eilbek 61:i:z	61
ser Eimsbuettel 6,(7),(14):d:l,w	C1
*ser Ejeda 45:a:z10	W
ser Ekotedo (9),46:z4,z23:-	D2
ser Elizabethville 3,10:r:1,7	E1
ser Elomrane 1,9,12:z38:-	D1
*ser Elsiesivier 16:z42:1,6	I
ser Emek (8),20:g,m,s:	C3
ser Emmastad 38:r:1,6	P
*ser Emmerich 6,14:[m,t]:e,n,x	H
ser Encino 1,6,14,25:d:l,z13,z28	H
ser Enschede 35:z10:l,w	O
ser Entebbe 1,4,12,27:z:z6	B
ser Enteritidis 1,9,12:g,m:-	D1
ser Enugu 16:l,z13,z28:-	I
ser Epicrates 3,10:b:l,w	E1
ser Eppendorm [1],4,12,[27]:d:1,5	B
*ser Epping 13,23:e,n,x:1,7	G2
*ser Erlangen 48:g,m,t:-	Y
ser Escanaba 6,7:k:e,n,z15	C1
ser Eschweiler 6,7:z10:1,6	C1
ser Essen 4,12:g,m:-	B
ser Etterbeek 11:z4,z23:e,n,z15	F
ser Exra 28:z:1,7	M
ser Fako 1,42:a:e,n,z15	T
ser Falkensee 3,10:i:e,n,z15	E1
ser Fallowfield 3,10:l,z13,z28:e,n,z15	E1
ser Fandran 1,40:z35:e,n,x,z15	R
ser Fann 11:l,v:e,n,x	F
ser Fanti 13,23:z38:-	G2

Organism	O Group
ser Farcha 43:y:1,2	U
*ser Farmsen 13,23:z:1,6	G2
*ser Faure 50:z42:1,7	Z
ser Fayed 6,8:l,w:1,2	C2
ser Ferlac 1,6,14,25:a:e,n,x	H
*ser Finchley 3,10:z:e,n,x	E1
ser Findorff 11:d:z6	F
ser Finkenwerder 1,6,14,25:d:1,5	H
ser Fischerhuette 16:a:e,n,z15	I
ser Fischerkietzi 6,14,25:y:e,n,x	H
ser Fischerstrasse 44:d:e,n,z15	V
ser Fitzroy 48:e,h:1,5	Y
*ser Flint 50:z4,z23:-	Z
ser Florida 1,6,14,25:y:e,n,x	H
ser Flottbek 52:b:-	52
ser Fluntern 6,14,18:b:1,5	H
ser Fortune 4,12,27:z10:z6	B
*ser Foulpointe 38:g,t:-	P
ser Frankfurt 16:e:e,n,z15	I
ser Freetown 38:y:1,5	P
*Fremantle 42:(f),g,t:-	T
ser Fresno (9),46:z38:-	D2
* variant (9),46:z38:-	D2
ser Friedenau 13,22:d:1,6	G1
ser Friedrichsfelde 28:f,g:-	M
ser Frintrop 1,9,12:b:1,5	D1
*ser Fuhlsbuettel 3,10:l,v:z6	E1
ser Fulica 4,5,12:a:1,5	B
ser Gabon 6,7:l,w:1,2	C1
ser Galiema 6,7:k:1,2	C1
ser Galil 3,10:a:e,n,z15	E1
bioser Gallinarum 1,9,12:-:-	D1
ser Gamaba 44:g,m,s:-	V
ser Gambaga 21:z35:e,n,z15	L
ser Gambia 35:i:e,n,z15	O
ser Gamanara 16:d:1,7	I
ser Garba 1,6,14,25:a:1,5	H
ser Garoli 6,7:i:1,6	C1
ser Gassi 35:e,h:z6	O
ser Gateshead (9),46:g,s,t:-	D2
ser Gatow 6,7:y:1,7	C1
ser Gatuni 6,8:b:e,n,x	C2
ser Gdansk 6,7:l,v:z6	C1
ser Gege 30:r:1,5	N
ser Gelsenkirchen 6,(7),(14):l,v:z6	C1
ser Georgia 6,7:b:e,n,z15	C1
ser Gera 1,42:z4,z23:1,6	T
*ser Germiston 6,8:m,t:e,n,x	C2
ser Ghana 21:b:1,6	L
ser Geissen 30:g,m,s:-	N

Organism	O Group
*ser Gilbert 6,7:z39:a,7	C1
ser Give 3,10:l,v:1,7	E1
ser Glasgow 16:b:1,6	I
*ser Glencairn 11:a:z6:z42	F
ser Glostrup 6,8:z10:e,n,z15	C2
ser Gloucester 1,4,12,(27):i:l,w	B
ser Gnesta 1,3,19:b:1,5	E4
ser Godesberg 30:g,m:	n
ser Goelzau 3,10:A:1,5	E1
ser Goerlitz 3,15:e,h:1,2	E2
ser Goeteborg 9,12:c:1,5	D1
ser Soettingen 9,12:l,v:e,n,z15	D1
*ser Gojenberg 1,13,23:g,t:1,5	G2
ser Gokul 1,51:d:-	51
ser Goldcoast 6,8:r:11,w	C2
ser Gombe 6,7:d:e,n,z15	C1
ser Good 21:f,g:e,n,x	L
*ser Goodwood 13,22:z29:e,n,x	G1
ser Gori 17:z:1,2	J
ser Goulfy 1,40:k:1,5,(6)	R
ser Goverdhan 9,12:k:1,6	D1
*ser Grabouw 11:g,m,s,t:[z39]	F
ser Graz 43:a:1,2	U
*ser Greenside 50:z:e,n,x	Z
ser Greiz 40:a:z6	R
ser Grumpensis 13,23:d:1,7	G2
*ser Grunty 1,40:z39:1,6	R
ser Guilford 28:k:1,2	M
ser Guinea 44:z10:[1,7]	V
*ser Gwaai 21:z4,z24:-	L
ser Gwoza 1,3,19:a:e,n,z15	E4
ser Haardt (8):k:1,5	C3
*ser Haarlem (9),46:z:e,n,x	D2
ser Habana = ser Havana	
ser Hadar 6,8:z10:e,n,x	C2
*ser Haddon 16:z4,z23:-	I
ser Haelsingborg 6,7:m,p,t,[u]	C1
ser Haferbreite 42:k:[1,6]	T
*ser Hagenbeck 48:d:z6	Y
ser Haifa 1,4,5,12:z10:1,2	B
variant afula 01 & 05- 4,12:z10:1,2	B
ser Halle 28a,28c:c:1,7	M
variant Vidin 28a,28b:c:1,7	M
ser Halmstad 3,15:g,s,t:-	E2
*ser Hamburg 1,9,12:g,t:-	E2
ser Hamilton 3,15:z27:-	
= R phase of ser Goerlitz	E2
*ser Hammonia 48:e,n,x,z15:z6	Y
ser Hannover 16:a:1,2	I
ser Haouari 13,22:c:e,n,x,z15	G1

Organism	O Group
ser Harburg 1,6,14,25:k:1,5	H
*ser Harmelen 51:z4,z23:-	51
ser Harrisonburg (3),(15),34:z10:1,6	E3
ser Hartford 6,7:y:e,n,x	C1
ser Harvestehude 1,42:y:z6	T
ser Hato 4,5,12:g,m,s:-	B
ser Havana 1,13,23:f,g,[s]:-	G2
ser Heerlen 11:i:1,6	F
ser Heidelberg [1],4,[5],12:r:1,2	B
*ser Heilbron 6,7:l,z28:1,5:[z42]	C1
*ser Helsinki 1,4,12:z29:[e,n,x]	B
*ser Hennepin 41:d:z6	S
ser Hermannswerder 28:c:1,5	M
ser Heron 16:a:z6	I
ser Herston 6,8:d:e,n,z15	C2
ser Herzliya 11:y:e,n,x	F
ser Hessarek 4,12,[27]:a:1,5	B
ser Heves 6,14,24:d:1,5	H
ser Hidalgo 6,8:e:e,n,z15	C2
*ser Hillbrow 17:b:e,n,x,z15	J
ser Hillegersberg (9),46:z35:1,5	D2
ser Hillsborough 6,7:z41:l,w	C1
ser Hilversum 30:k:2	N
ser Hindmarsh (8):r:1,5	C3
ser Hirschfeldii = ser Paratyphi C	
ser Hisingen 48:a:1,5,7	Y
ser Hofit 39:i:1,5	Q
ser Holcomb 6,8:l,v:e,n,x	C2
ser Holstein, not a salmonella	
ser Homosassa 1,6,14,25:z:1,5	H
ser Honelis 28:a:e,n,z15	M
*ser Hooggraven 50:z10:z6:z42	Z
ser Horsham 1,6,14,25:l,v:e,n,x	H
*ser Houten 43:z4,z23:-	U
*ser Hueningen 9,12:z:z39	D1
*ser Hila 11:l,z28:e,n,x	F
ser Hull 16:b:1,2	I
*ser Humber 53:z4,z24:-	53
ser Huvudsta 3,10:b:1,7	E1
ser Hvittingfoss 16:b:e,n,x	I
ser Ibadan 13:22:b:1,5	G1
ser Idikan 13,23:i:1,5	G2
ser Ilala 28:k:1,5	M
ser Illinois (3),(15),34:z10:1,5	E3
ser Illugun (1),3,10,(19):z4,z23:z6	E4
ser Indiana 1,4,12:z:1,7	B
ser Infantis 6,7,[14]:r:1,5	C1
ser Inganda 6,7:z10:1,5	C1
ser Inglis (9),46:z10:e,n,x	D2
ser Inpraw 41:z10:e,n,x	S

Organism	O Group
ser Inverness 38:k:1,6	P
ser Ipeko 9,12:c:1,6	D1
ser Ipswich 41:z4,z24:-	S
ser Irenea 17:k:1,5	J
ser Irigny 43:z38:-	U
ser Irumu 6,7:l,v:1,5	C1
*ser Islington 3,10:g,t:-	E1
ser Israel 9,12:e,h:e,n,z15	D1
ser Isuge 13,23:d:z6	G2
ser Italiana 9,12:l,v:1,11	D1
ser Ituri 1,4,12:z10:1,5	B
ser Itutabe (9),46:c:z6	D2
ser Iwojima - ser Kentucky	
*ser Jacksonville 16:z29:-	I
ser Jaffna 1,9,12:d:z35	D1
ser Jaja 4,12,27:z4,z23:-	B
ser Jamaica 9,12:r:1,5	D1
ser Jangwani 17:a:1,5	J
bioser Java 1,4,5,12:b:[1,2]	B
ser Javiana 1,9,12,27:c:e,n,z15	D1
ser Jedgurgh 3,10:z29:-	E1
ser Jericho 1,4,12,27:c:e,n,z15	B
ser Jerusalem 6,(7),[14]:z10:l,w	C1
ser Jodhpur 45:z29:-	W
ser Joenkoeping 4,5,12:g,s,t:-	B
ser Johannesburg 1,40:b:e,n,x	R
ser Jos 1,4,12,27:y:e,n,z15	B
ser Jukestown 13,23:i:e,n,z15	G2
ser Kaapstad 4,12:e,h:1,7	B
ser Kaduna 6,(7),(14):c:e,n,z15	C1
ser Kahla 1,42:z35:1,6	T
ser Kaitaan 1,6,14,25:m,t:-	H
ser Kalamu 1,4,12:z4,z24:[1,5]	B
ser Kalina 3,10:b:1,2	E1
*ser Kaltenhausen 28:b:z6	M
ser Kamoru 4,12,27:y:z6	B
ser Kampala 1,42:c:z6	T
ser Kanda = ser Meleagridis	
ser Kandla 17:z29:-	J
ser Kaneshie 1,42:i:l,w	T
ser Kaolack 47:z:1,6	X
ser Kapemba 9,12:l,v:1,7	D1
ser Kaposvar, combined w/ser Reading	
ser Karachi 45:d:e,n,x	W
ser Karamoja 40:z41:1,2	R
ser Kasenyi 38:e,h:1,5	P
ser Kassberg 1,6,14,25:c:1,6	H
*ser Katesgrove 1,13,23:m,t:1,5	G2
ser Kentucky (8),20:i:z6	C3
variant Jerusalem (8):i:z6	C2

Organism	O Group
ser Kenya 6,7:l,z13:e,n,x	C1
*ser Khami 47:b:e,n,x,z15	X
ser Khartoum (3),(15),34:a:1,7	E3
ser Kiambu 4,12:z:1,5	B
ser Kibi 16:z4,z23:-	I
ser Kibusi 28:r:e,n,x	M
ser Kidderminster 38:c:1,6	P
ser Kiel 1,2,12:g,p:-	A
ser Kikoma 16:y:e,n,x	I
*ser Kilwa 4,12:l,w:e,n,x	B
ser Kimberly 38:l,v:1,5	P
ser Kimuenza 1,4,12,27:g,s,t:-	B
ser Kingabwa 43:y:1,5	U
ser Kingston 1,4,12,27:g,s,t:-	B
variant Copenhagen 4,12:g,s,t:-	B
ser Kinondoni 17:a:e,n,x	J
ser Kinshasa 3,15:l,z13:1,5	E2
ser Kintambo 13,23:m,t:-	G2
ser Kirkee 17:b:1,2	J
ser kisangi 1,4,5,12:a:1,2	B
ser Kisarawe 11:k:e,n,x	F
ser Kitenge 28:y:e,n,x	M
ser Kivu 6,7:d:1,6	C1
*ser Klapmuts 45:z:z39	W
*ser Kluetjenfelde 4,12:d:e,n,x	B
ser Kokemlemle 39:l,v:e,n,x	Q
*ser Kommetje 43:b:z42	U
ser Korbol (8),20:b:1,5,(6)	C3
ser Korelbu 1,3,19:z:1,5	E4
ser Korovi 38:g,m,s:-	P
ser Kottbus 6,8:e,h:1,5	C2
ser Kotte 6,7:b:z35	C1
ser Koumra 6,7:b:1,7	C1
ser Kralendyk 6,7:z4,z24	C1
*ser Kraaifontein 1,13,23:g,(m),t:[e,n,x]	G2
ser Kralingen (80,20:y:z6	C3
ser Krefeld 1,3,19:y:l,w	E4
ser Kristianstad 3,10:z10:e,n,z15	E1
*ser Krugersdorp 50:e,n,x:1,7	Z
ser Kuessel 28:i:e,n,z15	M
*ser Kuilsrivier 1,9,12:g,m,s,t:e,n,x	D1
ser Kumasi 30:z10:e,n,z15	N
ser Kunduchi 1,4,[5],12,27:l,z28:1,2	B
ser Kuru 6,8:z:l,w	C2
ser Labadi 6,8:d:z6	C2
ser Lagos 1,4,12:i:1,5	B
ser Landala 41:z10:1,6	S
ser Landau 30:i:1,2	N
ser Langenhorn 18:m,t:-	K
ser Langensalza 3,10:y:l,w	E1

Organism	O Group
ser Langford 28:b:e,n,z15	M
ser Lanka 3,15:r:z6	E2
ser Lansing 38:i:1,5	P
ser Larochelle 6,7:e,h:1,2	C1
ser Lattenkamp 45:z35:1,5	W
ser Lawndale 1,9,12:x:1,5	D1
ser Lawra 44:k:e,n,z15	V
ser Leeuwarden 11:b:1,5	F
ser Legon [1],4,12,[27]:c:1,5	B
ser Leiden 13,22:z38:-	G1
ser Leipzig 41:z10:1,5	S
ser Leith 6,8:a:e,n,z15	C2
ser Lekke 3,10:d:1,6	E1
ser Leoben 28:l,v:1,5	M
ser Leopoldville 6,7:b:z6	C1
*ser Lethe 41:g,t:-	S
ser Lexington 3,10:z10:1,5	E1
ser Lezennes 6,8:z4,z23:1,7	C2
*ser Lichtenberg 41:z10:[z6]	S
ser Ligeo 30:l,v:1,2	N
ser Ligna 35:z10:z6	O
ser Lille 6,7:z38:-	C1
*ser Limbe 1,13,22:g,m,t:[1,5]	G1
ser Limete 1,4,12,27:b:1,5	B
*ser Lincoln 11:m,t:e,n,x	F
ser Lindenburg 6,8:i:1,2	C2
ser Lindern 6,14,24:d:e,n,x	H
ser Lindi 38:r:1,5	P
*ser Lindrick 9,12:e,n,x:l,[5],7	D1
ser Lingivala 16:z:1,7	I
ser Linton 13,23:r:e,n,z15	G2
ser Lisboa 16:z10:1,6	I
ser Lishabi (9),46:z10:1,7	D2
ser Litchfield 6,8:l,v:1,2	C2
ser Liverpool 1,3,19:d:e,n,z15	E4
ser Livingstone 6,7:d:l,w	C1
ser Ljubljana 4,12,27:k:e,n,x	B
ser Llandoff 1,3,19:z29:-	E4
*ser Llandudno 28:g,s,t:1,5	M
ser Loanda 6,8:l,v:1,5	C2
*ser Lobatsi 52:-:1,5,7	52
*ser Locarno 57:z29:z42	57
ser Loenga 1,42:z10:z6	T
ser Logone 39:d:1,5	Q
*ser Lohbruegge 44:z4,z32:-	V
ser Lokstedt 1,3,19:l,z13,z28:2	E4
ser Lomalinda 9,12:a:e,n,x	D1
ser Lome 9.12:r:z6	D1
ser Lomita 6,7:e,h:1,5	C1
ser London 3,10:l,v:1,6	E1

Organism	O Group
ser Losangeles 16:l,v:z6	I
ser Louga 30:b:1,2	N
*ser Louwbester 16:z:e,n,x	I
ser Lovelace 13,22:l,v:1,5	G1
*ser Luanshya 13,23:g,s,(t):-	G2
ser Luciana 11:a:e,n,z15	F
ser Luckenwalde 28:z10:e,n,z15	M
ser Luke 1,47:g,m:-	X
*ser Lundby (9),46:b:e,n,x	D2
*ser Lurup 41:z10:e,n,x,z15	S
*ser Luton 60:z:e,n,x	60
ser Lyon 47:k:e,n,z15	X
*ser Maarssen (9),46:z4,z24:z39:z42	D2
ser Maartensdyijk 40:g,p:-	R
ser Maastricht 11:z41:1,2	F
ser Macallen 3,10:d:e,n,z15	F
ser Machaga 1,3,19:i:e,n,x	E4
ser Madelia 1,6,14,25:y:1,7	H
ser Madiago 1,3,19:c:1,7	E4
ser Madigan 44:c:1,5	V
ser Madjorio 3,10:d:e,n,z15	E1
ser Magumeri 1,6,14,25:e,h:1,6	H
ser Magwa 21:d:e,n,x	L
ser Maiduguri 1,3,19:f,g,t:e,n,z15	E4
ser Makiso 6,7:l,z13,z28:z6	C1
*ser Makoma 4,12:a:-	B
*ser Makumira [1],4,12,[27]:e,n,x:1,7	B
ser Malakai 16:e,h:1,2	I
ser Malstatt 16:b:z6	I
ser Mampeza 1,6,14,25:i:1,5	H
ser Mampong 13,22:z35:1,6	G1
ser Manchester 6,8:l,v:1,7	C2
ser Mandera 16:l,z13:e,n,z15	I
ser Manhattan 6,8:d:1,5	C2
*ser Manica 1,9,12:g,m,s,t:z42	D1
ser Manila 3,15:z10:1,5	E2
*ser Manombo 57:z39:e,n,x,z15	57
ser Mapo 6,8:z10:1,5	C2
ser Mara 39:e,h:[1,5]	Q
ser Maracaibo 11:l,v:1,5	F
ser Maricopa 1,42:g,z51:1,5	T
ser Marienthal 3,10:k:e,n,z15	E1
*ser Marina 48:g,z51:-	Y
ser Maritza = ser Salford var Maritza	
ser Maron 3,10:d:z35	E1
ser Marseille 11:a:1,5	F
ser Marylebone (9),46:k:1,2	D2
ser Massakory 35:r:l,w	O
ser Massenya 1,4,12,27:k:1,5	B
ser Matadi 17:k:e,n,x	J

Organism	O Group
ser Mathura (9),46:i:e,n,z15	D2
ser Matopeni 30:y:1,2	N
*ser Matroosfontein 3,10:a:c,n,x	E1
ser Mayday (9),46:y:z6	D2
ser Mbandaka [1],6,7,[25]:z10:e,n,z15	C1
ser Mbao 43:i:1,2	U
ser Meleagridis 3,10:e,h:l,w	E1
ser Memphis 18:k:1,5	K
ser Menden 6,7:z10:1,2	E1
ser Mendoza 9,12:l,v:1,2	D1
ser Menhaden (3),(15),34:l,v:1,7	E3
ser Menston 6,7:g,s,t:1,5	C1
*ser Merseyside 16:,g,t:1,5	I
ser Mesbit 47:m,t:e,n,z15	X
ser Meskin 51:e,h:1,2	51
ser Messina 30:d:1,5	N
ser Mexicana, combined w/ser Muenchen	
ser Mgulani 38:i:1,2	P
ser Miami 1,9,12:a:1,5	D1
ser Michigan 17:l,v:1,5	J
ser Middlesbrough 1,42:i:z6	T
*ser Midhurst 53:l,z28:z39	53
ser Mikawasima 6,7:y:e,n,z15	C1
ser Millesi 1,40:l,v:1,2	R
ser Milwaukee 43:f,g:-	U
ser Mim 13,22:a:1,6	G1
ser Minneapolis (3),(15),34:e,h:1,6	E3
ser Minnesota 21:b:e,n,x	L
ser Mishmarhaemek 1,13,23:d.1,5	G2
ser Mission 6,7:d:1,5	C1
ser Mississippi (Atlanta) 1,13,23:b:1,5	G2
ser Missouri 11:g,s,t:-	F
ser Miyazaki 9,12:b:e,n,x	D1
*ser Mjimwema 1,9,12:b:e,n,x	D1
*ser Mobeni 16:g,m,s,t:-	I
ser Mocamedes 28:d:e,n,x	M
ser Moero 28:b:1,5	M
ser Mokola 3,10:y:1,7	E1
ser Molade (8),20:z10:z6	C3
*ser Mondeor 39:l,z28:e,n,x	Q
ser Mons 1,4,12,[27]:d:l,w	B
ser Monschaui 35:m,t:-	O
ser Montevideo 6,7:g,m,s:-	C1
*ser Montgomery 11:d,a:d,e,n,z15	F
ser Montreal = ser Vein	
ser Morehead 30:i:1,5	N
ser Morocco 30:l,z13,z28:e,n,z15	N
ser Morotai 17:l,v:1,2	J
ser Moroto 28:z10:l,w	M
ser Moscow 9,12:g,q:-	D1

Organism	O Group
*ser Mosselbay 43:g,s,(t):z42	U
ser Moualine 47:y:1,6	X
ser Mountpleasant 47:z:1,5	X
ser Mowanjum 6,8:z:1,5	C2
*ser Mpila 3,10:z38:z42	E1
ser Muenchen 6,8:d:1,2	C2
ser Muenster 3,10:e,h:1,5	E1
ser Muguga 44:m,t:-	V
*ser Muizenberg 9,12:g,m,s,t:1,5	D
ser Mundonobo 28:d:1,7	M
*ser Mundsburg 11:g,z51:-	F
ser Mura 1,4,12:z10:i,w	B
*ser Maachshonim 1,13,23:z:1,5	G2
ser Maestved 1,9,12:g,p,s:-	D1
ser Nagoya 6,8:b:1,5	C2
*ser Mairobi 42:r:-	T
ser Nakura 1,4,12,27:a:z6	B
*ser Namib 50:g,m,s,t:1,5	Z
ser Napoli 1,9,12:l,z13:e,n,x	D1
ser Narashino 6,8:a:e,n,x	C2
ser Nashua 28:l,v:e,n,z15	M
ser Nchanga 3,10:l,v:1,2	E1
ser Ndolo [1],9,12:d:1,5	D1
*ser Neasden 9,12:g,s,t:e,n,x	D1
*ser Negev 41:z10:1,2	S
ser Nessa 1,6,14,25:z10:1,5	H
ser Nessziona 6,7:l,z13:1,5	C1
ser Neukoelln 6,7:l,z13,z28:e,n,z15	C1
ser Neumuenster 1,4,12,27:k:1,6	B
* variant 1,4,12,27:k:1,6	B
ser Newbrunswick 3,15:l,v:1,7	E2
ser Newhaw 3,15:e,h:1,5	E2
ser Newington 3,15:e,h:1,6	E2
ser Newlands 3,10:e,h:e,n,x	E1
ser Newmexico 9,12:g,z51:1,5	D1
ser Newport 6,8:e,h:1,2	C2
ser Newrochelle 3,10:k:l,w	E1
ser Newyork = ser Javiana	
ser Ngili 6,7:z10:1,7	C1
ser Ngor 1,3,19:l,v:1,5	E4
*ser Ngosi 48:z10:[1,5]	Y
ser Niamey 17:d:l,w	J
ser Niarembe 44:a:l,w	V
ser Nienstedten 6,(7),(14):b:[l,w]	C1
ser Nieukerk 6,(7),(14):d:z6	C1
variant zollenspicker 6,7:d:z6	C1
ser Nigeria 6,7:r:1,6	C1
ser Nikolaifleet 16:g,m,s:-	I
ser Niloese 1,3,19:d:z6	E4
ser Nima 28:y:1,5	M

Organism	O Group
ser Nipponbasi, not confirmed	
ser Nissii 6,7,14:b:-	C1
ser Nitra 1,12:g,m:-	A
*ser Noordhoek 16:l,w:z6	I
*ser Nordenham 1,4,12,27:z:e,n,x	B
ser Nordufer 6,8:a:1,7	C2
ser Norton 6,7:i:l,w	C1
ser Norwich 6,7:e,h:1,6	C1
ser Nottingham 16:d:e,n,z15	I
ser Nowawes 40:z:z6	R
ser Nuatja 16:d:e,n,z15	I
*ser Nuernberg 42:z:z6	T
ser Nyanza 11:z:z6	F
ser Nyborg 3,10:e,h:1,7	E1
ser Oahu, not confirmed	
ser Oakland 6,7:z:1,6,(7)	C1
ser Obogu 6,7:z4,z23:1,5	C1
ser Ochsenwerder 54:k:1,5	54
*ser Ochsenzoll 16:z4,z23:-	I
*ser Odijk 30:a:z39	N
ser Odozi 30:k:e,n,x,z15	N
* variant 30:k:e,n,x,z15	N
*ser Oevelgoenne 28:r:e,n,z15	M
ser offa 41:z38:-	S
ser Ohio 6,7:b:l,w	C1
ser Ohlstedt 3,10:y:e,n,x	E1
ser Okatie 13,23:g,s,t:-	G2
ser Okefoko 3,10:c:z6	E1
ser Okerara 3,10:z10:1,2	E1
ser Oldengurg 16:d:1,2	I
ser Omderman 6,(7),(145):d:e,n,x	C1
ser Omifisan 40:z29:-	R
ser Ona 28:g,s,t:-	M
ser Onarimon 1,9,12:b:1,2	E1
ser Onderstepoort 1,6,14,25:e,h:1,5	H
ser Onireke 3,10:d:1,7	E1
ser Oranienburg 6,7:m,t:-	C1
ser Ordonez 1,13,23,37:y:l,w	G2
ser Oregon, combined w/ser Muenchen	
ser Orientalis 16:k:e,n,z15	I
ser Orion 3,10:y:1,5	E1
ser Oritamerin 6,7:i:1,5	C1
ser Os 9,12:a:1,6	D1
ser Oskarshamn 28:y:1,2	M
ser Oslo 6,7:a:e,n,x	C1
ser Osnabrueck 11:l,z13,z28:e,n,x	F
ser Othmarschen 6,7:g,m,[t]:-	C1
*ser Ottershaw 40:d:-	R
ser Ouakam (9),[12],[34]:46:z29:-	D2
ser Overschie 51:l,v:1,5	51

Organism	O Group
ser Overvecht 30:a:1,2	N
ser Oxford 3,10:a:1,7	E1
*ser Oysterbeds 6,7:z:z47	C1
ser Pakistan (8):l,v:1,2	C3
ser Panamy 1,9,12:l,v:1,5	D1
ser Pankow 3,15:d:1,5	E2
ser Papuana 6.7:r:e,n,z15	C1
ser Paratyphi = ser Paratyphi A	
bioser Paratyphi A 1,2,12:a:-	A
variant Durazzo 2,12:a:-	A
ser Paratyphi B java = ser Java	
1,4,5,12:b:[1,2]	B
ser Paratyphi B 1,4,5,12:b:1,2	B
variant Odense 1,4,12:b:1,2	B
bioser Paratyphi C 6,7,[Vi]:c:1,5	C1
*ser Patera 11:z4,z23:-	F
*ser Parow 3,15:g,m,s,t:-	E2
ser Paris (8):20,z10:1,5	C3
ser Patience 28:d:e,n,z15	M
ser Penarth 9,12:z35,z6	D1
ser Pensacola 9,12:m,t:-	D1
*ser Perinet 45:m,t:e,n,x,z15	W
ser Perth 38:y:e,n,x	P
ser Pharr 11:b:e,n,z15	F
*ser Phoenix 47:b:1,5	X
ser Pikine (88),20:r;z6	C3
ser Plymouth (9),46:d:z6	D2
ser Poano 1,6,14,25:z:l,z13,z28	H
ser Poeseldorf 54:i:z6	54
ser Pomona 28:y:1,7	M
ser Poona [1],13,22,[37]:z:1,6	G1
variant 37 1,13,22,36,37:z:1,6	G1
*ser Portbech 42:l,v:e,n,x,z15	T
ser Portland 9,12:z10:1,5	D1
ser Portsmouth 3,15:l,v:1,6	E2
ser Potsdam 6,7:l,v:e,n,z15	C1
ser Potto (9),12,46:i:z6	D2
ser Phaha 6,8:y:e,n,z15	D2
ser Pramiso 3,10:c:1,7	E1
ser Presov 6,8:b:e,n,z15	C2
ser Preston 1,4,12:z:l,w	B
ser Pretoria 11:k:1,2	F
ser Pueris, combined w/ser Newport	
bioser Pullorum 9,12:-:-	D1
ser Pumila 48:a:z6	Y
ser Putten 13,23:d:l,w	G2
*ser Quimbamba 47:d:z39	X
ser Quinhon 47:z44:-	X
ser Quiniela 6,8:c:e,n,z15	C2
ser Ramatgen 30:k:1,5	N

Organism	O Group
*ser Rand 42:z:e,n,x,z15	T
ser Raus 13,22:f,g:e,n,x	G1
ser Reading 4,[5],12:e,h:1,5	B
ser Rechovot (8),20:e,h:z6	C3
ser Redhill 11:e,h:l,z13,z28	F
ser Redlands 16:z10:e,n,z15	I
ser Regent 3,10:f,g:-	E1
ser Remo 1,4,12,27:r:1,7	B
*ser Rhodensiense 9,12:d:e,n,x	D1
ser Rhone 21:c:e,n,x	L
ser Richmond 6,7:y:1,2	C1
ser Rideau 1,3,19:f,g:-	E4
ser Ridge 9,12:c:z6	D1
ser Riggil 6,7:g,t:-	C1
ser Riogrande 40:b:1,5	R
ser Rissen 6,7:f,g:-C1	C1
ser Riverside 45:b:1,5	W
ser Roan 38:l,v:e,n,x	P
ser Rochdale 50:b:e,n,x	Z
*ser Roggeveld 51:-:1,7	51
ser Rogy 28:z10:1,2	M
ser Romanby 13,23:z4,z24:-	G2
ser Roodepoort [1],13,22,[37]:z10:1,5	G1
*ser Rooikrantz1,6,14:m,t:1,5	H
ser Rosenthal 3,15:b:1,5	E2
ser Rossleben 54:e,h:1,6	54
ser Rostock 1,9,12:g,p,u:-	D1
*ser Roterberg 6,7:z4,z23:-	C1
*ser Rotterdam 1,13,22:g,t:1,5	G1
*ser Rowbarton 16:m,t:-	I
ser Rubislaw 11:[d],r:[d],e,n,x	F
ser Ruiru 21:y:e,n,x	L
ser Ruki 4,5,12:y:e,n,x	B
ser Rutgers 3,10:l,v:e,n,z15	E1
ser Ruzizi 3,10:l,v:e,n,z15	E1
ser Saarbruecken [1],9,12:a:1,7	D1
*ser Sachsenwald 1,40:z4,z23:-	R
ser Saintmarie 52:g,t:-	52
ser Saintpaul 1,4,[5],12:e,h:1,2	B
ser Saipam, not confirmed	
ser Saka 47:b:-	X
ser Sakai = ser Potsdam	
*ser Sakaraha 48:[k]:z39	Y
ser Salford 16:l,v:e,n,x	I
ser Salinatis 4,12:d,e,h:d,e,n,z15	B
ser Sandiego 4,12:e,h:e,n,z15	B
ser Sandow 6,8:f,g:e,n,z15	C2
ser Sanga (8):b:1,7	C3
ser Sanjuan 6,7:a:1,5	C1
ser Sanktgeorg 28:r,s(i):e,n,z15	M

Organism	O Group
ser Sanktmarx 1,3,19:e,h:1,7	E4
ser Santhiaba 40:l,z28:1,6	R
ser Sao 1,3,19:e,h:e,n,z15	E4
ser Saphra 16:y:1,5	I
ser Sara 1,6,14,25:z38:[e,n,x]	H
ser Sarajane 4,12,27:d:e,n,x	B
*ser Sarepta 16:l,z28:z42	I
ser Schalkwijk 6,14,(24):i:e,n....	H
ser Schleissheim 4,12,27:b,z12:-	B
ser Schoeneberg 1,3,19:z:e,n,z15	E4
ser Schottmuelleri = ser Paratyphi B	
ser Schwarzengrund 1,4,12,27:d:1,7	B
ser Schwerin 6,8:k:e,n,x	C2
*ser Seaforth 50:k:z6	Z
ser Seattle 28:a:e,n,x	M
ser Sedgwick 44:b:e,n,z15	V
ser Seegefeld 3,10:r,(i):1,2	E1
ser Sekondi 3,10:e,h:z6	E1
ser Selandia 3,15:e,h:1,7	E2
*ser Seminole 1,40a,40b:g:z51	R
bioser Sendai 1,9,12:a:1,5	D1
ser Senegal 11:r:1,5	F
ser Senftenberg 1,3,19:g,s,t:-	E4
ser Seremban 9,12:i:1,5	D1
*ser Setubal 60:g,m,t:z6	60
ser Shamba 16:c:e,n,x	I
ser Shangani 3,10:d:1,5	E1
ser Shanghai 16:l,v:1,6	I
ser Sharon 11:k:1,6	F
ser Sheffield 38:c:1,5	P
ser Shikmonah 40:a:1,5	R
ser Shipley (8),20:b:e,n,z15	C3
ser Shomolu 28:y:l,w	M
*ser Shomron 18:z4,z32:-	K
ser Shoreditch (9),46:r:e,n,z15	D2
ser Shubra 4,5,12:z:1,2	B
ser Siegburg 6,14,18:z4,z23:[1,5]	K
ser Simi 3,10:r:e,n,z15	E1
*ser Simonstown 1,6,14:z10:1,5	H
ser Simsbury 1,3,19:z27:-	E4
ser Singapore 6,7:k:e,n,x	C1
ser Sinstorf 3,10:l,v:1,5	E1
ser Sinthia 18:z38:-	K
ser Sladun 1,4,12,27:b:e,n,x	B
*ser Slangkop 1,6,14:z10:z6:z42	H
*ser Slatograd 30:f,g,(p),t:-	N
ser Sljeme 1,47:f,g:-	X
ser Sloterdijk 1,4,12,27:z35:z6	B
ser Soahamina 6,14,24:z:e,n,x	H
ser Soerenga 30:i:l,w	N

Organism	O Group
*ser Soesterberg 21:z4,z23:-	L
*ser Sofia 4,12,[27]:b:[e,n,x]	B
ser Solna 28:a:1,5	M
ser Solt 11:y:1,5	F
ser Southbank 3,10:m,t:-	E1
*ser Soutpan 11:z:z39	F
ser Souza 3,10:d:e,n,x	D1
ser Spartel 21:d:1,5	L
*ser Springs 40:a:z39	R
*ser Srinagar 11:b:e,n,x	F
ser Stanley 4,5,12:d:1,2	B
ser Stanleyville 1,4,5,12:z4,z23:[1,2]	B
ser Steinplatz 30:y:1,6	N
*ser Stellenbosch 1,9,12:z:1,7	D1
ser Stellingen 47:d:e,n,x	X
ser Stendal 11:l,v:1,2	F
ser Sternchanze 30:g,s,t:-	N
ser Sterrenbos 6,8:d:e,n,x	C2
*ser Stevenage 1,13,23:[z42]:1,7	G2
*ser Stikland 3,10:m,t:e,n,x	E1
ser Stockholm 3,10:y:z6	E1
ser Stormont 3,10:d:1,2	E1
ser Stourbridge 6,8:b:1,6	C2
ser Straengnaes 11:z10:1,5	F
ser Strasbourg (9),46:d:1,7	D2
ser Stratfprd 1,3,19:i:1,2	E4
*ser Suarez 1,40:c:e,n,x,z15	R
ser Suberu 3,10:g,m:	E
*ser Suederelbe 1,9,12:b:z39	D1
ser Sueliforf 45:f,g:-	W
ser Suez = ser Shubra	
ser Suipestifer = ser Choleraesuis	
*ser Sullivan 6,7:z42:1,7	C1
ser Sundsvall 1,6,14,25:z:e,n,x	H
ser Sunnycove (8):y:e,n,x	C3
*ser Sunnydale 1,40:k:e,n,x,z15	R
ser Surat 1,6,14,25:r,(i):e,n,z15	H
variant Hr- 1,6,14,25:i:e,n,z15	H
*ser Sydney 48:i:z	Y
ser Szentes 16:k:1,2	I
*ser Tafelbaai 3,10:z:z39	E1
ser Tafo 1,4,12,27:z35:1,7	B
ser Taihoku = ser Meleagridis	
ser Takorida 6,8:i:1,5	C2
ser Taksony 1,3,19:i:z6	E4
ser Tallahassee 6,8:z4,z32:-	C2
ser Tamale (8),20:z29:-	C3
ser Tananarive 6,8:y:1,5	C2
ser Tanger 1,13,22:y:1,6	G1
ser Tarshyne 9,12:d:1,6	D1

Organism	O Group
ser Taunton 28:k:e,n,x	M
ser Tchad 35:b:-	O
ser Techimani 28:c:z6	M
ser Teddington 4,12,27:y:1,7	B
ser Tees 16:f,g:-	I
ser Tejas 4,12:z36:-	B
ser Teko 1,6,14,25:d:e,n,z15	H
ser Telaviv 28:y:e,n,z15	M
ser Telekebir 13,23:d:e,n,z15	G2
ser Telhashomer 11:z10:e,n,x	F
ser Teltow 28:z4,z23:1,6	M
ser Tennessee 6,7:z29:-	C1
ser Teshie 1,47:l,z13,z28:e,n,z15	X
ser Texas 4,5,12:k:e,n,z15	B
ser Thiaroye 38:e,h:1,2	P
ser Thielallee 6,(7),(14):m,t:-	C1
ser Thomasville (3),(15),34:y:1,5	E3
ser Thompson 6,7,[14]:k:1,5	C1
variant Berlin 6,7:-:1,5	C1
variant 14 6,7,14:k:1,5	C1
ser Tilburg 1,3,19:d:l,w	E4
ser Tilene 1,40:e,h:1,2	R
ser Tim = ser Newington var tim	
ser Tinda 1,4,12,27:a:e,n,z15	B
ser Tione 51:a:e,n,x	51
ser Togo 4,12:ll,w:1,6	B
*ser Tokai 57:z42:1,6:z53	57
ser Tokoin 4,12:z10:e,n,z15	B
ser Tokyo, not confirmed	
ser Toney 54:b:e,n,x	54
ser Tornow 45:g,m:-	W
*ser tosamanga 6,7:z:1,5	C1
ser tournai 3,15:y:z6	E2
*ser Tranoroa 55:k:z39	55
ser Travis 4,5,12:g,z51:1,7	B
ser Treforest 1,51:z:1,6	51
ser Trotha 40:z10:z6	R
ser Tshiongwe 6,8:e,h:e,n,z15	C2
ser Tucson 1,6,14,25:b:[1,7}	H
ser Tuda 4,12:z10:1,6	B
ser Tuebingen 3,15:y:1,2	E2
*ser Tuindorp43:z4,z32:-	U
*ser Tulear 6,8:a:z52	C2
ser Tunis 1,13,23:y:z6	G2
*ser Tygerberg 1,13,23:a:z42	G2
Salmonella typhi 9,12,[Vi]:d:-	D1
S. enteritidis	
ser Typhimurium 1,4,5,12:i:1,2	B
variant binns 1,4,5,12:-:1,2	B
variant Copenhagen 1,4,12:i:1,2	B

Organism	O Group
bioser Typhisuis 6,7:[c]:1,5	C1
ser Ucclc 54:g,s,t:-	54
ser Uganda 3,10:l,z13:1,5	E1
ser Ughelli 3,10:r;1,5	E1
ser Uhlenhorst 44:z:l,w	V
ser Ullevi 1,13,23,27:b:e,n,x	G2
ser Umbilo 28:z10:e,n,x	M
ser Umhlali 6,7:a:1,6	C1
ser Umhlatazana 35:a:e,n,z15	O
*Unnamed serotypes	
ser _____4,12:(f),g:-	B
ser _____4,12:-:1,6	B
ser _____6,7:a:z6	C1
ser _____6,7:g,t:e,n,x:z42	C1
ser _____6,7:k:[z6]	C1
ser _____6,7:z:z6	C1
ser _____6,7:z10:z35	C1
ser_____6,7:z29:-	C1
ser _____6,7:z42:e,n,x:1,6	C1
ser _____6,8:g,(m),t:e,n,x	C2
ser _____9,12:e,n,x:1,6	D1
ser _____(9),46:z10:z6	D2
ser _____1,9,12,(46),27:y:z39	D2
ser _____3,10:l,z28,z39	E1
ser _____11:b:1,7	F
ser _____11:z4,z23:-	F
ser _____13,23:l,z28:z6	G2
ser _____ (6),14:k:[e,n,x]	H
ser _____1,(6),14:k:z6:z42	H
ser _____1,(6),14:z42:1,6	H
ser _____16:b:z42	I
ser _____16:l,z40:-	I
ser _____17:k:-	
ser _____18:b:1,5	K
ser _____18:m,t:1,5	K
ser _____18:y:e,n,x,z15	K
ser _____28:c,n,c:1,7	M
ser _____30:z39:1,(7)	N
ser _____35:g,m,s,t:-	O
ser _____35:l,z28:-	O
ser _____40:b:-	R
ser _____1,40a,40c:g,z51:-	R
ser _____1,40:m,t:z42	R
ser _____1,40:z6:1,5	R
ser _____1,40:-:1,7	R
ser _____41:k:-	S
ser _____42:m,t:e,n,x,z15	T
ser _____42:-:1,6	T
ser _____43:e,n,x,z15:l,(5),7	U
ser _____43:e,n,z15:1,6	U

Organism	O Group
ser _____43:z:1,5	U
ser _____44:g,z51:-	V
ser _____44:z4,z23:-	V
ser _____44:z36,z38:-	V
ser _____45:g,z51:-	W
ser _____47:z6:1,6	X
ser _____48:z4,z32:-	Y
ser _____50:l,w:e,n,x,z15:z42	Z
ser _____50:l,z28:z42	Z
ser _____50:z4,z24:-	Z
ser _____52:d:e,n,x,z15	52
ser _____53:z:z6	53
ser _____56:e,n,x:1,7	56
ser _____57:g,m,s,t:z42	57
ser _____58:a:-	58
ser _____58:a:1,5	58
ser _____64:k:e,n,x,z15	64
ser _____64:z29:-	64
ser Uno 6,8:z29	C2
*ser Uphill 42:b:e,n,x,z15	T
ser Uppsala 4,12,27:b:1,7	B
ser Urbana 30:b:e,n,x	N
ser Ursenbach 1,42:z:1,6	T
ser Usumbura 18:d:1,7	K
ser Utah 6,8:c:1,5	C2
*ser Utbremen 35:z29:e,n,x	O
ser Utrecht 52:d:1,5	52
ser Uzaramo 1,6,14,25:z4,z24:-	H
ser Vaertan 13,22:b:e,n,x	G1
ser Vancouver 16:c:1,5	I
*ser Veddel 43:g,t:-	U
ser Vejle 3,10:e,h:1,2	E1
ser Vellore 1,4,12,27:z10:z35	B
ser Venusberg = ser Nchanga variant venusberg	
ser Veneziana 11:e:e,n,x	F
*ser Verity 17:e,n,x,z15:1,6	J
ser Victoria 1,9,12:l,w:1,5	D1
ser Victoriaborg 17:c:1,6	J
ser Vietnam 41:b:-	S
* variant 41:b:-	S
ser Vinohrady 28:m,t:-	M
* variant 28:m,t:-	M
ser Virchow 6,7:r:1,2	C1
ser Virginia (8):d:[1,2]	C3
ser Visby 1,3,19:b:1,6	E4
ser Vitkin 28:l,v:e,n,x	M
ser Vleuten 44:f,g:-	V
ser Volkmarsdorf 28:i:1,6	M
*ser Volksdorf 43:z36,z38:-	U
ser Volta 11:r:l,z13,z28:-	F

Organism	O Group
ser Vom 4,12,27:l,z13,z28:e,n,z15	B
*ser Vredelust 1,13,23:l,z28:z42	B2
*ser Vrindaban 45:a:e,n,x	W
ser Wa 16:b:1,5	I
ser Wagenia 1,4,12,27:b:e,n,z15	B
*ser Wandsbek 21:z10:z6	L
ser Wandsworth 39:b:1,2	Q
ser Wangata 9,12:z4,z23:[1,7]	D1
ser Warnow 6,8:i:1,6	C2
ser Warragul 1,6,14,25:g,m:-	H
*ser Wassenaar 50:g,z51:-	Z
ser Waycross 41:z4,z23:-	S
* variant 41:z4,z23:-	S
ser Wayne 30:g,z51:-	N
ser Wedding 28:c:e,n,z15	M
ser Welikade 16:l,v:1,7	I
ser Weltevreden 3,10:r:z6	E1
ser Wentworth 11:z10:1,2	F
ser Wernigerode (9),46:f,g:-	D2
ser Weslaco 42:z36:-	T
ser Westerstede 1,3,19:l,z13:-	E4
ser Westhampton 3,10:g,s,t:-	E1
ser Weston 16:e,n:z6	I
*ser Westpark 3,10:l,z28:e,n,x	E1
ser Westphalia 35:z4,z24:-	O
ser Weybridge 3,10:d:z6	E1
ser Wichita 1,13,23:d:z37	G2
ser Wien 1,4,12,[27]:b:l,w	B
ser Wil 6,7:d:z13,z28	C1
ser Wildwood (3),(15),34:e,n:l,w	E3
ser Wilhelmsburg 4,[5],12,[27]:z38:-	B
variant Teufelsbrueck 1,4,12:z38:-	B
*ser Wilhelmstrasse 52:z44:1,5,7	52
ser Willemstad 1,13,22:e,h:1,6	G1
ser Wimborne 3,10:z39:1,7	E1
*ser Winchester 3,10:z39:1,7	E1
ser Windermere 39:y:1,5	Q
*ser Windhoek 45:g,t:1,5	W
ser Wingrove 6,8:c:1,2	C2
ser Wippra 6,8:z10:z6	C2
*ser Woerden 17:c:z39	J
ser Womba 4,12,27:c:1,7	B
*ser Woodstock 16:z42:l,(5),7	I
*ser Worcester 1,13,23:z:l,w	G2
ser Worthington 1,13,23:z:l,w	G2
*ser Wynberg 1,9,12:z39:1,7	D1
ser Yaba 3,10:b:e,n,z15	*E1*
ser Yalding 1,3,19:r:e,n,z15	E4
ser Yarm 6,8:z35:1,2	*C2*
ser Yarrabah 13,23:y:1,7	G2

Organism	O Group
ser Yeerongpilly 3,10:i:z6	E1
ser Yerba 54:z4,z23:-	54
ser Yodabasi, not confirmed	
ser Yoff 38:z4,z23:1,2	P
ser Yolo 35:c:-	O
ser Zadar (9),46:b:1,6	D2
ser Zagreb, combined wser Saintpaul	
ser Zanzibar 3,10:k:1,5	E1
ser Zega 9,12:d:z6	D1
ser Zehlendorf 30:a:1,5	N
*ser Zeist 18:z10:z6	K
*ser Zuerich 1,9,12:c:z39	D1
ser Zuilen 1,3,19:i:l,w	E4
ser Zwickau 16:r,(i):e,n,z15	I

*   = Biochemically aberrant.
( ) = Antigen incomplete.
[ ] = Antigen may be present or absent.

REFERENCES

1. **Difco Laboratories,** *Difco Manual*, 10th ed., Difco Laboratories, Detroit, 1984.
2. **Edwards, P. R. and Ewing, W. H.,** *Identification of Enterobacteriaceae*, 3rd ed., Burgess Publishing Company, Minneapolis, MN, 1972.

Appendix C

REAGENTS USEFUL IN MICROBIOLOGICAL STUDY

Brilliant Green Dye Solution

Dissolve 1 g brilliant green dye in 100 ml sterile distilled water. Always test each lot of dye for toxicity before use in tests.

Brilliant Green Water

Make up a solution by adding 2 ml of a 1% brilliant green dye solution to 1000 ml distilled water.

Bromcresol Purple Dye Solution

Dissolve 0.2 g of bromcresol purple dye in 100 ml of sterile distilled water. Always test each lot of dye for toxicity before use in tests.

Buffered Saline

Phosphate buffered saline is a most useful diluent and suspending fluid for washing samples of microorganisms from the surfaces of meats. It may be prepared as follows:

NaCl	8.00 g/l
K_2HPO_4	1.21 g/l
KH_2PO_4	0.34 g/l

This solution gives a pH of approximately 7.3 when prepared in distilled water and provides both the potassium and phosphate ions for microorganisms. Other types of buffers may be used for other purposes.

Earle's Balanced Salts (Phenol Red-Free)

NaCl	6.8 g
KCl	400 mg
$CaCl_2.2H_2O$	265 mg
$MgSO_4.7H_2O$	200 mg
$NaH_2PO_4.H_2O$	140 mg
Glucose	1.0 g
$NaHCO_3$	2.2 g
Distilled water	1 liter

Dissolve ingredients in water and sterilize by filtration. Final pH should be 7.2.

Formalinized Physiological Saline

Formaldehyde (37%)	6 ml
NaCl	8.5 g
Distilled water	1 liter

Dissolve sodium chloride in water and autoclave for 15 min at 121°C. When cool, add the formaldehyde. Do not autoclave after formaldehyde addition.

Gram Stain
Commercial staining solutions are satisfactory. If it is desired to mix fresh in the laboratory, the following formulations are satisfactory.

Gram I. Crystal or Gentian Violet — 2 g. Dissolve in 95% ethanol — 20 ml. Dissolve 0.8 g ammonium oxalate in distilled water — 80 ml. Mix the two solutions. Allow to stand for 24 h and filter through coarse filter paper.

Gram II. Iodine — 1 g. Potassium iodide (KI) — 2 g. May be necessary to mix iodine and KI in pestle and add water slowly with grinding between additions. When dissolved, pour into reagent bottle and rinse mortar and pestle with water as needed to bring volume to 300 ml total.

Gram III. Ethanol, 95%.

Gram IV. Counterstain. Dissolve 2.5 g Safranin O (certified) in 100 ml 95% ethanol. Add 10 ml of this stock dye to 90 ml distilled water for use.

For staining — fix microbial smear with moderate heat. *DO NOT OVERHEAT*. Stain for approximately 1 min by flooding the smear with Gram I. Wash with tap water. Flood with Gram II for approximately 1 min. Wash with Gram III until no more color comes off (approximately 20 to 30 s). Wash with tap water. Flood with Gram IV for approximately 1 min. Wash with tap water and blot dry.

Hydrochloric Acid (1 N)
Concentrated HCl — 89 ml. Add slowly, with mixing to water to make total volume of 1000 ml.

Kovacs' Reagent
Used to detect indole production.

Amyl or isoamyl alcohol	150 ml
p-Dimethylaminobenzaldehyde	10 g
Hydrochloric acid, concentrated	50 ml

Dissolve the aldehyde in the alcohol and then slowly add the acid. The reagent should be light colored. If a deep brown color results, discard and use a different alcohol. The reagent is stable at room temperature. It is recommended that large amounts not be prepared at one time, and that the reagent be stored at refrigerator temperature when not in use.

To perform the test, add 0.5 ml of the reagent to a 48 h culture which has been incubated at 37°C. Shake the tube gently. A deep-red color develops in the presence of indole.

McFarland Barium Sulfate Standards (McFarland Nephelometer)

These standards are sealed tubes of barium sulfate suspensions prepared to use as standards in adjusting the densities of bacterial suspensions. It is obvious that the diameter of the tubes in which the suspensions are measured must be comparable to that of the standard. Thorough mixing of the standard each time it is used is necessary. While visual comparisons may be made, more accurate measurements are made by use of electronic nephelometers. The McFarland standards are prepared in a ten tube series from a 1% solution of barium chloride in distilled water, and a 1% solution of sulfuric acid in distilled water. The solutions are mixed in the series as listed in the table below, and the corresponding density of bacterial cells in millions per milliliter are also listed.

Tube number	Barium chloride (ml)	Sulfuric acid (ml)	Cell density (millions/ml)
1	0.1	9.9	300
2	0.2	9.8	600
3	0.3	9.7	900
4	0.4	9.6	1200
5	0.5	9.5	1500
6	0.6	9.4	1800
7	0.7	9.3	2100
8	0.8	9.2	2400
9	0.9	9.1	2700
10	1.0	9.0	3000

Methyl Red Test Indicator

Prepare the solution of methyl red in 95% ethyl alcohol by adding 0.1 g methyl red to 300 ml of alcohol. Add water to the solution to make a total of 500 ml.

For the test, use a 48 h culture in glucose-peptone broth which has been incubated at 37°C. Add 5 drops of reagent for each 5 ml of culture. Positive tests are bright red. A weak positive may be red-orange, and negative tests are yellow.

Phosphate Buffers

Phosphate buffers may be made up as stock solutions to be mixed as needed in larger amounts. It is convenient to prepare 1000 ml of stock solutions which can then be mixed according to the following table to prepare a buffer of the desired pH.

Solution A is a 0.2 M solution of monobasic sodium phosphate prepared by dissolving 31.2 g of NaH_2PO_4, $2H_2O$ in 1000 ml of distilled water. Solution B is a 0.2 M solution of dibasic sodium phosphate prepared by dissolving 28.39 g of Na_2HPO_4 or 71.7 g of $Na_2HPO_4.12H_2O$ in 1000 ml of distilled water.

Buffer for use is prepared in 200 ml increments according to the table following:

Solution A (ml)	Solution B (ml)	Distilled water (ml)	Buffer pH
92.0	8.0	100	5.8
87.7	12.3	100	6.0
81.5	18.5	100	6.2
73.5	26.5	100	6.4
62.5	37.5	100	6.6
51.0	49.0	100	6.8
39.0	61.0	100	7.0
28.0	72.0	100	7.2
19.0	81.0	100	7.4
13.0	87.0	100	7.6
8.5	91.5	100	7.8
5.3	94.7	100	8.0

pH Indicator Dyes

A table of indicators and the ranges of pH producing color changes follows:

Indicator	pH Range	Color change
Thymol blue (acid range)	1.2–2.8	Red to yellow
Bromphenol blue	2.8–4.6	Yellow to violet
Bromcresol green	3.6–5.2	Yellow to blue
Methyl red	4.4–6.2	Red to yellow
Litmus	4.5–8.3	Red to blue
Bromcresol purple	5.2–6.8	Yellow to violet
Bromthymol blue	6.0–7.6	Yellow to blue
Neutral red	6.8–8.0	Red to yellow
Phenol red pink	6.8–8.4	Yellow to purple
Cresol red	7.2–8.8	Yellow to violet-red
Thymol blue (alk range)	8.0–9.6	Yellow to blue
Phenolphthalein	8.3–10.0	Colorless to red
Thymolphthalein	9.3–10.5	Colorless to blue

Physiological Saline

This is an NaCl solution in distilled water to approximate the osmotic pressure of mammalian blood serum. It is prepared by solution of 0.85 g of NaCl in 100 ml of water.

Potassium Hydroxide Solution (40%)

Add 40 g of KOH to water to make a final volume of 100 ml.

Sodium Hydroxide (1 N)

Add 40 g of NaOH to water to make one liter.

Tergitol Anionic 7

This reagent is manufactured by Union Carbide Corp., Chemicals and Plastics, 270 Park Avenue, New York, NY 10017. It is recommended for use as a wetting and emulsifying agent when the electrolyte is below 1% in textiles etc. It is a sodium sulfate derivative of 3,9-diethyl tridecanol-6. It is used as a surfactant in preparation of coconut and some meats for culture for *Salmonella*.

Sorensen Buffer Solutions

Solution A: M/15 Na_2PO_4
   Dissolve 9.464 g of the anhydrous salt to make 1 liter.
Solution B: M/15 KH_2PO_4
   Dissolve 9.073 g of the anhydrous salt to make 1 liter.
Mix Solution A and Solution B according to the table below to attain the pH indicated.

pH	Solution A (ml)	Solution B (ml)
5.29	0.25	9.75
5.59	0.5	9.5
5.91	1	9
6.24	2	8
6.47	3	7
6.64	4	6
6.81	5	5
6.98	6	4
7.17	7	3
7.38	8	2
7.73	9	1
8.04	9.5	0.5

Multiply the above amounts by increments of 10 to achieve the desired volume of buffer at any pH. These buffers are also useful in some uses to combine with NaCl for phosphate buffered saline.

Triton X-100
This reagent is manufactured by Rhom and Haas Co., Independence Mall West, Philadelphia, PA 19105. It is a nonionic preparation of octylphenoxy polyethyoxy ethanol, used as a wetting agent and dispersant or emulsifier. It is needed in preparation of coconut and some meats for culture for *Salmonella*

Voges-Proskauer (VP) Test Reagents
More than one test has been described for detection of the production of acetylmethylcarbinol as an end product of glucose metabolism by enteric bacteria. The reagents listed here are those used in the Barritt modification of the test.

Solution A
  alpha-Napthol   5 g
  Ethyl Alcohol, absolute   100 ml

Solution B
  Potassium hydroxide   40 g
  Distilled water   100 ml

The test is performed by adding 0.6 ml of A and 0.2 ml of B to 1 ml of culture. Shake well after the addition of each reagent. Positive reactions, development of a red color, will occur within 5 min. A copper color development is not a positive reaction, and should be discarded.

REFERENCES FOR OTHER REAGENTS AND METHODS

1. **Baron, E. J. and Finegold, S. M.,** *Bailey & Scott's Diagnostic Microbiology,* 10th ed., C. V. Mosby Company, St. Louis, 1990.
2. **Food and Drug Administration,** *Bacteriological Analytical Manual,* Association of Official Analytical Chemists, Arlington, VA, 1984.
3. **Harrigan, W. F. and McCance, M. E.,** *Laboratory Methods in Food and Dairy Microbiology,* Revised ed., Academic Press, New York, 1990.

GLOSSARY

Adjuvant — A compound that aids, facilitates, or enhances the functioning of another substance.

Adulterate — To make impure by the addition of a foreign, inferior, or harmful substance.

Antisepsis — The use of antiseptics.

Antiseptic — A chemical substance that destroys or inhibits the action of microorganisms on living tissues, having the effect of limiting or preventing the harmful effects of infection.

Agglutination — Clumping of bacterial cells by specific antiserum.

Antibiotic — Destructive of life. A chemical substance produced by a microorganism which has the capacity to inhibit the growth of or to kill other microorganisms.

Allochthonous — Foreign. Not naturally occurring. Transient.

Antibody — An immunoglobulin molecule that has a specific amino acid sequence by virtue of which it interacts only with the antigen that induced its synthesis.

Antibody titer — The highest dilution of a serum which visibly reacts with a specific antigen.

Antigenic mosaic — The pattern of different antigenic molecules on the surface of a bacterial cell.

Asymptomatic — Showing or causing no symptoms.

Autochthonous — Indigenous.

Bacteremia — Presence of bacteria in the blood.

Bactericide — A chemical that destroys or inhibits the growth of bacteria.

Bacteriophage — A virus that lyses bacteria.

Bacteriophage conversion — Carriage of gene or genetic material by a bacteriophage to a bacterial cell during infection.

Bacteriostasis — The inhibition of growth, but not the killing, of bacterial cells.

Bacteriostatic agent — A chemical which accomplishes bacteriostasis.

Beta-lactamase — An enzyme produced by some bacteria which hydrolyzes some penicillins.

Bradycardia — Slowness of heartbeat, as evidenced by slowing of the pulse rate to less than 60.

Carrier — An individual who harbors in his body the specific organisms of a disease without manifest symptoms and thus acts as a carrier or distributor of the infection.

Catalysis — Increase in the velocity of a chemical reaction or process produced by the presence of a substance that is not consumed in the net chemical reaction or process.

Chemotherapeutic agent — A chemical used to treat an infection that affects the causative organism unfavorably but does not harm the patient.

Chloramphenicol acetyltransferase — An enzyme produced by some bacteria which inactivates the drug chloramphenicol.

Cholecystectomy — Surgical removal of the gallbladder.

Clean or cleanse — To free from dirt, pollution, or foreign substance.

Colon — That part of the large intestine which extends from the cecum to the rectum.

Composite — Made up of distinct parts. Sample make-up when preparing foods for test.

Conjugation — The one-way transfer of DNA between bacteria in cellular contact.

Convalescent — A patient who is recovering from a disease.

Coproantibodies — Antibodies present in the intestinal tract associated with immunity to enteric infections.

Critical control point — A step or point in an operation at which an effective preventive or control measure can be exercised.

Cross-resistance — Resistance of a bacterium to an antibiotic substance related to one which has been used to treat a clinical condition.

Detergent — A cleansing agent which aids in emulsification of oils and removal of dirt or soils from a surface.

Determinant group — That portion of a chemical molecule, usually protein, which ascribes to the molecule its antigenic specificity.

Dehydration — The condition which results from excessive loss of body water.

Diarrhea — Abnormal frequency and liquidity of fecal discharges.

Differential medium — A culture medium which allows the growth of some bacteria and gives them distinguishing characteristics to help separate them from others in a mixed culture.

Disinfect — To free from infection, especially the destruction of harmful microorganisms.

Disinfectant — A chemical agent which causes the destruction or killing of microorganisms capable of causing infection.

Ecological niche, spatial niche — Spatial, physical habitat with interactions of organisms within that space.

Emulsifer — A surface-active agent (as a soap) which promotes and stabilizes the formation of an emulsion.

Endemic — Present in a community at all times.

Endotoxin — A heat-stable toxin present in the bacterial cell but not in cell-free filtrates of cultures of intact bacteria.

Enrichment — Culture in a medium which allows some bacteria to grow while others are slowed or inhibited.

Enteritis — Inflammation of small intestine.

Enterocolitis — Inflammation of small intestine and colon.

Enterotoxin — A toxin specific for the cells of the intestinal mucosa. One of the factors in production of diarrhea.

Episome — Any accessory, extrachromosomal, replicating, genetic element that can exist either autonomously or integrated with the chromosome.

Epithelium — The covering of internal and external surfaces of the body, including the lining of vessels and other small cavities.

Exotoxin — A soluble toxin molecule produced by a microorganism and released from the cell of that organism.

Focal infections — Points at which infections occur. Organs in which infections occur.

Fungicide — A chemical agent that destroys fungi or inhibits their growth.

Gastroenteritis — Inflammation of the stomach and intestine.

Germicide — A vague term which should be avoided. Technically a chemical which kills germs.

Good Manufacturing Practice (GMP) — A specific regulation dealing with a step or an operation in the processing of foods.

Gram-negative — Bacteria which fail to retain the primary dye (gentian-violet) in a staining procedure devised by Gram for microorganisms.

Gram-positive — Bacteria which retain the primary dye (gentian-violet) in a staining procedure devised by Gram for microorganisms.

Habitat — Physical location where an organism naturally exists.

HACCP (Hazard Analysis and Critical Control Point) — A system of inspection and control for food production, processing, or service operations.

Hard water — Water that contains alkaline metal ions, mainly calcium and magnesium, at a concentration above 60 ppm. These ions react with soaps and interfere with the emulsifying activity of the soap.

Hazard — An unacceptable contamination, growth, or survival of microorganisms which might cause illness to the consumer, or spoilage of food.

Hepatitis — Inflammation of the liver.

Ileum — The portion of the small intestine extending from the jejunum to the cecum.

Immunization — Inoculation with a specific antigen to induce an immune response.

Incubation period — The period of induction of an infectious disease after contact with the disease organism.

Indicator bacteria — Those bacteria most often found in large numbers in one location (i.e., in animal feces), which when found in a different location indicate pollution.

Indigenous — Naturally occurring. Native.

Infection — The state produced by the establishment of an infectious agent in or on a susceptible host. Also an infectious or contagious disease.

Infective dose, infectious dose — That number of infectious agent particles necessary to initiate an infection in a normal host.

Inflammatory lesions — Lesions in tissues resulting from inflammation.

In vitro — Observable in a test tube.

In vivo — Observable in a living body.

K antigens — Antigens present in the capsule of a bacterium. In the *Salmonella* these antigens are designated Vi antigens.

Kauffmann-White scheme or Kauffmann-White schema — A system of classification of *Salmonella* bacteria by determination of the presence of somatic (O), flagellar (H), and virulence (Vi) antigens in an organism. The antigens are listed, rather than the organisms being given a species or strain name.

Krad — Radiation dose absorbed ×1000 in reference to treatment of foods.

Leukocyte count — The number of leucokytes, or white blood cells, present in a standard volume of blood.

Lymphatics — Pertaining to lymph tissue or lymph vessels.

Lysogenic — Pertaining to lysogenicity.

Lysogenicity, lysogeny — The specific association of the phage genome, the prophage, with the bacterial genome in such a way that few, if any, phage genes are transcribed.

Lysogenization — Becoming lysogenic.

Mesenteric — Pertaining to the mesentery — a membranous fold attaching various organs to the body wall.

Minimum infective dose — The smallest number of bacterial cells which are capable of initiating an infection.

Monoclonal antibodies — Antibodies produced by a single cell, and therefore, possessing greater specificity.

Monocyte — A mononuclear, phagocytic leukocyte.

Monocytic — Pertaining to the monocyte.

Mortality rate — The rate of deaths in a population.

Mucosa — A mucous membrane within the body.

Myalgia — Pain in a muscle or muscles.

Necrosis — Death of tissue, usually as individual cells, groups of cells, or in small localized areas.

Nucleic acid — Any of the various acids composed of a sugar or derivative of a sugar, phosphoric acid, and a base, and found especially in the nucleus of a cell.

Osteomyelitis — Inflammation of a bone caused by a pyogenic organism.

Pasteurization — The process of heating milk, or other liquids, to a moderate temperature for a definite time (62°C for 30 min, or 71.7°C for 15 s, or 90°C for 0.5 s). This process kills the vegetative cells, but not the spores, of most pathogenic bacteria.

Pathogen — A specific causative agent of disease.

Pathogenesis — The origination and development of a disease.

Peptize — To cause proteins to become colloidal.

Peritoneum — The serous membrane lining the abdominopelvic walls and investing the viscera.

Peyer's Patches — Plaques on lymphatic tissues caused by typhoid infection.

Phase variation — A completely reversible, immunological variation most often in the bacterial flagella.

Phenol coefficient — The dilution of a chemical or disinfectant which will kill a standard bacterial suspension in a given time as compared to the dilution of phenol required to kill in the same time.

Plasmid — A generic term for all types of intracellular inclusions that can be considered to have genetic function.

Pollute — To make impure or unclean.

Polymorphonuclear leukocyte — A leukocyte having a nucleus so deeply lobed or so divided that it appears to be multiple.

Potable water — Water suitable for drinking. Does not contain harmful chemicals or organisms.

Pre-enrichment — Inoculation of sample into a nonselective medium for incubation prior to inoculation into selective medium.

Production lot — All units of a product of one size, produced under essentially the same conditions at one plant within a measured time span.

Prognosis — A forecast as to the probable outcome of an attack of a disease.

Prophage — An intracellular form of a bacteriophage in which it is harmless to the host. It usually is integrated into the hereditary material of the host, and reproduces when the host does.

Pure culture — A bacterial culture which contains only one species or type of organism.

Purine — A colorless, crystalline, heterocyclic compound which is not found free in nature but which is variously substituted to produce a group of compounds known as purine bases (purine bodies), of which uric acid is a metabolic end product.

Rad — Radiation dose absorbed. Sometimes referred to as rep in reference to treatment of foods.

Reactive site — That portion of an antibody protein molecule which ascribes specificity for a certain antigen.

Recall — The removal of all of one lot of food product from the market by order of regulatory authorities.

R factor — Plasmids which carry resistance genes for one or more antibiotics.

Ribosome — Intracellular, ribonucleoprotein particles concerned with protein synthesis.

Rose spots — Rash lesions of typhoid fever.

S-R dissociation — The change from a smooth, or encapsulated, to a rough, or nonencapsulated form in bacteria. This change is frequently spontaneous, and is the result of genetic alterations within the cells.

Sanitary — Harmless to human health and well being. Of good quality.

Selective medium — A culture medium which allows the growth of some bacteria while inhibiting the growth of others.

Septicemia — Systemic disease with the presence and persistence of pathogenic microorganisms or their toxins in the blood. Usually implies that the organisms are multiplying within the blood, or are being spilled into the blood as they multiply.

Serovar — A variety of organisms classified on the basis of the antigens present in the structure.

Soft water — Water that contains metallic ions such as calcium and magnesium in a concentration of less than 60 ppm.

Subtherapeutic — An applied dose of drug which is less than that needed to treat an infection.

Sterilize — To remove all forms of life from an environment. To kill all forms of life in an environment.

Temperate virus — A virus which on infection, produces lysogeny in the host cell.

Thoracic duct — The large lymph duct which ascends from the peritoneal cavity to the junction of the left subclavian and the left jugular vein.

Toxin — A chemical produced by living organisms which is poisonous to man or animals.

Tranduction — The transfer of genetic determinants from one microorganism to another by a viral agent (bacteriophage).

Water activity — The ratio of water vapor pressure of a food to that of pure water at the same temperature. A measure of water availability to microorganisms present.

White blood cell count — The count or number of white blood cells or leukocytes in a standard volume of blood.

Zoonoses — Diseases of animals transmissible to man.

Zoonotic — Pertaining to zoonosis, or zoonoses.

INDEX

number reported, 94
hirschfeldii, 33
infantis
  antibiotic resistance, 96
  number reported, 94
krefeld, 91
london, 36
montevideo, 94
newport
  antibiotic resistance, 91, 96–97
  number reported, 94
nima, 43
panama, 33, 36
paratyphi
  clinical spectrum, 41
  host restriction, serovars with, 64
  phage typing, 33
  subtypes, 33
  taxonomy, 29, 31
pullorum, 117, 123, 149, 150
  infective doses, 45
  microbiological test considerations, 151
reading, 94
saint-paul, 91, 96
san diego, 96
schottmuelleri, 33
senftenberg, 94
Salmonella-Shigella (SS) agar, 147, 149
Salmonella
  thompson, 94
  typhi, see also Typhoid
    carrier states, 2–3
    host restriction, serovars with, 64
    phage typing, 33
    phenol coefficient determination, 83–84
    taxonomy, 29, 31
    WHO classification, 55
  typhimurium, 36, 117
    antibiotic resistance, 38–39, 91, 95–96
    incidence of, 94–95
    number reported, 94
    phage typing, 33
    selective media, 151
    waterborne, 12
  typhisuis, 55, 149–151
Salmonellosis
  clinical spectrum, see Clinical spectrum
  incidence of, 1, 14
Sample size, food processing controls, 133–134
Sanitation practices, 49, 114, see also
    Cleaning agents; Controls

egg handling, 123
poultry incidence, 123
Selective enrichment
  media preparation methods, 157–168
  methods, 136, 140, 146–149
Selenite broth, 146
Selenite cystine broth, 165–166
Selenite lysine agar (SLA), 139
Sensitivity testing, 87
Septicemia, 56
  clinical spectrum, 42
  typhoid fever, 41
Serotypes/serotyping, 27, 32–33, 152, see
    also Taxonomy
  antibiotic sensitivity and resistance, 93–96
  antiserum absorption, 30
  host restriction, 64
  Kaufmann-White scheme, 24, 28–32, 169–194
  methods, 136–137, 151–154
  nontyphoid salmonellosis, 55
  Salmonella enteritidis in eggs, 117, 118, 120–122
  typhoid, 41–42, 44–46
Serratia marcescens, 26
Sewage, 18, 19
Sexual reproduction, 27
Sheep, 36, 55
Shellfish, 13, 18, 19
Shigella, 18, 23, 147–148
Simmons citrate agar, 166
Slide agglutination tests, 32–33, 154
Smooth-rough dissociation, 32
Sodium hydroxide, 72, 199
Somatic antigens, 28–32, see also O
    antigens
Sorensen buffer solutions, 199
Soy flour, 143
Spicer-Edwards flagellar test, 150
Spices, 13, 145
Spread plate method, 134–135
Staining, 27
Staphylococcus aureus, 83–84
Strains, see Serotypes/serotyping;
    Taxonomy
Streptomycin
  mode of action, 85
  resistant strains, 90, 91, 96
Sucrose, 148, 149
Sulfadiazine resistance, 90, 91
Sulfamethoxazole, mode of action, 85, 86
Sulfapyratide, 148